Construction Equipment

2/758   15.99

Management of Construction Equipment

Frank Harris
Professor and Head, Dept of Civil Engineering
Loughborough University of Technology

Ronald McCaffer
Professor and Head, School of Construction, Engineering and Technology
Wolverhampton Polytechnic

Second Edition

MACMILLAN

First edition published 1982 by
GRANADA PUBLISHING LTD under the title
Construction Plant: Management and Investment Decisions

Second edition published 1991 by
MACMILLAN EDUCATION LTD
Houndmills, Basingstoke, Hampshire RG21 2XS
and London
Companies and representatives
throughout the world

Printed in Hong Kong

British Library Cataloguing in Publication Data
Harris, Frank
Management of construction equipment
1. Construction industries. Management
I. Title II. McCaffer, Ronald
624.068
ISBN 0–333–52727–5

Contents

Preface	vi
Acknowledgements	viii
Introduction	ix

Section 1: Objectives, policies and strategies — **1**
1 Equipment policy — 3
2 Hire and rental of equipment — 12
3 Organisation of hire companies and departments — 21

Section 2: Investment and procurement — **27**
4 Economic comparisons of equipment alternatives — 29
5 Equipment profitability — 60
6 Equipment acquisition — 80
7 Systematic selection of equipment — 91
8 Calculating a hire rate — 101

Section 3: Operational management — **117**
9 Maintenance of equipment — 119
10 Health and safety at work: regulations and requirements — 144
11 Insurance and licensing legalities — 166

Section 4: Financial and budgetary control — **185**
12 Budgetary control and costing — 187
13 Cash flow — 199
14 Financial management — 227
15 Using computers in managing construction equipment — 248
16 Overseas operations — 257

Appendix: Interest and time relationships and tabulations for interest rates of 10% and 15% — 264
Index — 273

Preface

Technical improvements in construction plant and equipment have caused a movement away from manual labour towards mechanisation on construction projects. This has raised the potential productivity of construction workers, while also necessitating relatively large capital investments in machinery and equipment on site which must be operated at an economic level of utilisation if an adequate rate of return on the capital employed is to be achieved. The old option of hiring and firing labour to suit the work load has disappeared. As a consequence, increasing numbers of firms, especially those engaged in heavy civil engineering or specialised building, are now placing far more emphasis on the selection, performance monitoring, control and maintenance of their equipment fleets.

At the same time independent equipment hire businesses and DIY small tools outlets have rapidly expanded to satisfy the short-term equipment needs of construction companies. The plant hire and equipment rental sectors are still in their formative phases, but already approximately 50% of all equipment used in the construction industry is hired or rented.

Because hiring and small tools renting is a relatively recent phenomenon, detailed and documented analyses describing the nature of construction equipment management and its associated organisations are few. During recent years, however, the authors have made a special study of the management of construction equipment and the results of their investigations have been incorporated into both undergraduate and postgraduate teaching. This has naturally directed attention towards a textbook on the subject, as it is essential that students receive guidance on the management of construction equipment at the onset of their careers, and this book has therefore evolved to meet the needs of engineers and builders whose roles are expanding to embrace the increasingly important management of construction equipment. Specifically this second edition covers both the management of equipment within a construction company and the independent hire or rental firm, and deals with policies, strategies and organisational structures from the small to the large concern. By necessity, emphasis is placed on the financial aspects of equipment acquisition and control, and modern capital investment decision-making techniques are given special consideration. Operational management is dealt with fully, including health and safety, licensing and insurance and the problems faced in overseas work: in particular, the logistic problems of maintaining construction equipment are discussed in detail. Finally, the book brings together the recent developments in computer technology and its application to improving equipment management.

This book will be beneficial to students and practitioners of civil engineering, construction and building. Specifically, it will find a market among contractors, hire and rental firms, engineers, builders, quantity surveyors, specification writers, equipment manufacturers, project managers and insurance and legal advisers. It will also serve as a useful reference for equipment managers in the industry.

Wolverhampton and Loughborough, 1991   F. C. H.
                                                          R. McC.

Acknowledgements

The authors are particularly grateful to Tony Kettle, an accountant in private practice, for advice on some aspects of financial management; Michael Ellis, Loughborough University Safety Officer, for guidance and comments on the Health and Safety at Work Act; and Peter Harlow, Head of Information at the Chartered Institute of Building, for suggesting many excellent reference sources. The book has also benefited from the unstinting efforts of Tony Thorpe and Pat Carrillo in locating the diffused published material on the subjects covered and checking much of the text. Thanks are also extended to accountants and insurance underwriters for their useful contributions, who, for professional reasons, must remain anonymous. The authors are also grateful to former students who have assisted in extending the body of knowledge in the subject of Management. Finally, the authors wish to thank Vera Cole, who cheerfully performed the task of typing the manuscript, and friends and colleagues, who helped in the preparation of the artwork.

Introduction

For many years there has been an underlying trend towards a greater use of equipment in construction. The scale of modern construction work, and the short construction times required, make the extensive use of equipment essential. Furthermore, as the costs of labour have increased, so have the benefits of using more machinery, and at site level this has provided the opportunity to achieve greater output per employee.

Unlike firms engaged in labour-intensive construction work and with no equipment holdings, plant owners require substantial and continuing capital investment to generate turnover. The profit/turnover ratio of construction companies is usually very low (e.g. 1–3%), while the turnover/capital employed ratio is high. The turnover could be 8–15 times the capital employed. For example: if the profit/turnover were 3% and the turnover/capital employed were 10, the profit/capital employed would be 30%. A large fleet of construction equipment would disturb this relationship by increasing the capital employed, thereby reducing the turnover/capital employed ratio. In these circumstances high utilisation of equipment is necessary to generate adequate revenue and a sufficient return on capital.

Consequently, civil engineering and building contractors who hold large equipment fleets must give very careful consideration to equipment selection, the method of acquisition, the monitoring of usage and performance, and the maintenance of their fleets. Thus, a separate management organisation has developed around the ownership and use of equipment.

The independent hire sector has expanded as the trend towards using more mechanical equipment for construction has developed. It now comprises some 5 000 companies, accounting for about 50% of all the equipment used. Equipment hire companies serve two purposes: they not only provide for hire specialised equipment which no single construction company could expect to utilise fully, but also hold and hire everyday items of equipment, thereby relieving construction companies of the need to own and manage their own fleets. As a result, equipment hire firms have acquired specific management skills which are peculiar to owning and operating equipment. This has stimulated them to operate their equipment more efficiently and perhaps more profitably than a construction company holding a smaller fleet. Thus, the need for equipment management is divided between construction companies who have their own equipment department or division and the hire companies including rental shops. The body of knowledge described as equipment management comprises management organisation, economic evaluations, budgetary control and costing, cash flow and financial management. It also involves maintenance and control of maintenance costs, the use of computers particularly for financial and costing purposes, and knowledge of health and safety laws, road transport laws and insurance. This book covers these subjects in four sections:

Section 1: objectives, policies and strategies

This section deals with the objectives of external equipment hire and rental firms, and the internal plant or equipment department of a construction company operating in profit- and service-centre markets, respectively.

Section 2: investment and procurement

This section deals with the economic criteria for evaluating invstments in equipment, including the effects of corporation tax, capital allowances and inflation. These economic analyses are also applied to the various forms of ownership. A detailed procedure for selecting the most suitable equipment for acquisition from a range of available alternatives is described, and the calculation of an economic hire rate taking account of ownership costs is given in a detailed example.

Section 3: operational management

The various strategies available for effective control of equipment maintenance and its costs are described in this section. The Health and Safety at Work Act and the requirements for inspection and testing of plant and equipment are summarised. Legal and contractual insurance requirements, especially hired plant and equipment are dealt with, and the various licences for plant operation on the public roads are reviewed.

Section 4: financial and budgetary control

This section deals with budgetary control and costing, cash flow and financial management as applied to hire companies and equipment divisions. The use of computers for these applications is explained. A chapter is also included commenting on the problems of operations in overseas conditions.

Section 1

Objectives, Policies and Strategies

Chapter 1
Equipment Policy

1.1 INTRODUCTION

The development of the plant holding in a construction company's organisation is often the natural result of previous construction activities. Equipment items are gradually acquired to service contracts until the holding eventually reaches a size where it becomes necessary to organise the fleet into a separate unit. This marks the beginning of a plant division or even subsidiary, which may continue to develop, supplying the needs of the parent company. In some cases it may become a fully independent concern.

Equipment ownership is not a fundamental requirement for a construction company, as a vast selection of plant is available for hire as an alternative. Many contractors, however, choose to own some of the equipment they need, for reasons of convenience and prestige. The decision to acquire equipment for profit should be considered carefully: unlike construction work itself, plant should be a capital-intensive business and requires a relatively large central organisation to provide all the facilities for maintenance, cost accounting, hiring, etc. Furthermore, an appropriate strategy for achieving profitability for a plant division may not always coincide with that of the construction division.

Therefore, it is a major management task to lay down principal policies for the supply and organisation of equipment holdings associated with a company's construction activities, so that the objectives for all the company's operations may be achieved. The consequence of these policies will stand as a record for later revision as required. If this is not done, the plant fleet may become a hidden and ever-increasing drain on financial resources.

1.2   EQUIPMENT ACQUISITION POLICY

1.2.1   Acquisition Options (Figure 1.1)

The means of obtaining construction plant may be broadly classified as follows:

(1) Owning all equipment – including hire purchase, straight purchase and leasing.
(2) Hiring all equipment.
(3) Combining owning and hiring equipment.

Each method will make special demands on the use of the company's capital and resources.

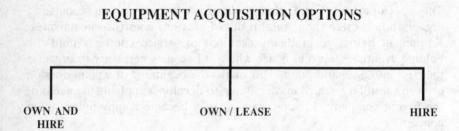

EQUIPMENT ACQUISITION OPTIONS

OWN AND HIRE   OWN / LEASE   HIRE

Figure 1.1   Plant acquisition options

Owning All Equipment

The policy practised by many construction companies is to purchase most of the equipment required. Plant availability is maintained thereby at all times, with the added advantage of the prestige attached to demonstrating the use of owned equipment. However, much capital will be locked up in the plant, which must become capable of generating a sufficient rate of return. A major disadvantage of owning a large plant fleet is the problem of maintaining adequate levels of utilisation. Equipment holdings are usually built up to service a growing demand for construction work, and will become a heavy liability when an economic recession occurs. The plant operator could be forced to seek any available work to sustain the fleet, since equipment cannot easily be sold in a declining market.

Hiring All Equipment

Many specialist plant hire firms are available for the supply of plant and equipment. The contractor who takes advantage of this facility avoids both the responsibility for maintenance and care of the equipment and tying up capital. The equipment may be rented for a specified period and hire charges minimised by standing off-hire all unwanted items. In many instances the plant operative is also provided by the equipment supplier.

   The main disadvantage of hiring is that hire rates depend on market forces and are largely beyond the control of the contractor, except for limited negotiation between competing firms. This vulnerability to changes in the industry's economic climate could seriously affect a contractor's quoted price for work and the costs incurred later when the work is carried out.

Combining Owning and Hiring Equipment

A company may prefer a mixed policy of owning and hiring plant. For example, equipment that is required for work on most contracts may be purchased and hiring adopted only to smooth out demand.

1.2.2   General Policy

Setting the Objectives

(1) The Type and Quality of Service to Offer

The management of a construction company, which has included as one of its main objectives the necessity to own and operate equipment rather than hire, must organise itself to either (1) Provide plant to service the company's contracts at rates of hire which compare reasonably with those available in the open market; or (2) Operate the plant division as a separate entity responsible for generating its own capital and profits, with the freedom to hire equipment to clients outside the parent company.

(2) The Type and Amount of Service to Offer

For the provision of plant as a 'service' arrangement, management must decide how much of the firm's capital should be invested in equipment, thereby setting indirectly the limits on the proportion of the firm's plant needs that will be self-owned. For the 'profit centre' system of operation, the plant subsidiary is an independent organisation, and a specific share of the market for equipment hire must be established as a major goal. Both systems will involve taking decisions on the appropriate selections of equipment.

*(3) The Possible Changes and Fluctuations in the Market for
Construction Work*

The fortunes of the construction industry will fluctuate with the needs of
the national economy. For example, there may be a decline in the road
building programme or an expansion of the offshore oil industry. These
changes will affect the demand for construction plant and equipment, and
management should always be looking ahead at the potential consequences.

Finally, whether the plant holdings are organised into a 'profit' or
'service' centre, it must be profitable. Clearly, there will be more opportunities for an independent plant company which can set its own objectives
and decide on strategies, but even a small service plant department should
maintain commercial viability.

1.2.3 Operational Policy

The logical way to operate an equipment holding is to make it an
independent profit centre. This applies particularly to a company trading
solely on the basis of hiring equipment in the open market. However,
many construction companies have plant holdings, some quite large, which
are not subject to open competition, and it is important that the capital
invested be used efficiently. For those firms where the proportion of
capital invested in equipment is high, it is obligatory to try to maximise the
profit on the investment; otherwise the capital would be better used
elsewhere in the business. Any other approach will carry with it serious
dangers. For example, if equipment is purchased to provide a service to
other parts of the construction company and charged below market rates,
the resulting low bids must lead ultimately to other parts of the business
having to earn excessive profits to generate an acceptable rate of return on
the total capital employed.

Several studies have shown that plant operated as a profit centre
generally out-performs plant as a service centre as far as profitability is
concerned. In the case of the former, the rigours of the market place
ensure that only equipment which can show a high level of utilisation
and/or profit throughout its working life is purchased. In addition, the
constraints are such that the costs of maintenance are controlled so that
neither too much nor too little is undertaken. For the service centre
structure, however, operating costs are ultimately met by the company as a
whole, and there is a tendency both for maintenance facilities to expand
and items to be purchased, with little regard to levels of utilisation. The
several organisational arrangements described below have evolved to
accommodate the needs of the company, each having consequences which
affect the efficient use of resources.

1.2.4   Alternative Options for Equipment Supply (Figure 1.2)

An Independent Hirer or Rental Shop

This type of organisation covers both the independent plant hire company including DIY small tools outlets, and the plant hire subsidiary operating under the umbrella of a holding company which may or may not be a construction firm. The company will supply equipment to the market to provide a satisfactory rate of return on the employed capital, although sometimes discounts may be offered to a parent company, in accordance with a policy towards a 'favoured' or important client. Decisions on equipment purchases and plant holding policy will be taken by the subsidiary board, with only the major objectives set by the parent company.

Normally, the subsidiary company will be given a name which is not associated with the parent firm, so that other construction companies will not be discouraged by advertising a competitor's name on a construction site.

Plant for the hire fleet or equipment shop will be purchased to make a profit based on its utilisation potential and the maintenance record experienced with similar items, together with an assessment of the hire rates likely to prevail in the future.

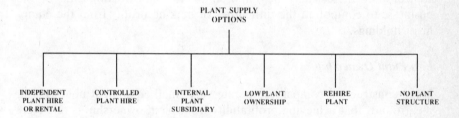

Figure 1.2   Plant supply options

Controlled Plant

The plant fleet of a construction company may ultimately become so extensive that, to maintain effective control, the holdings are incorporated into a subsidiary division. The first priority may be to serve the equipment needs of the parent company at a profit, but in order to maintain high levels of plant utilisation and thereby maximise profits, items may be hired out to other users. Many construction companies recommend as a rule of thumb that the ratio between 'hiring internally to the parent company' and 'hiring externally to the market' should not be less than 2:1.

However, the tendency with this system is for the equipment to be hired out to the market when the rates are attractive, since there will be a demand for such services and the required utilisation levels can be more easily achieved this way. Consequently, there is a danger that the needs of the parent company may be neglected, and items will not be available at the right time for the company's own construction contracts. Furthermore, the servicing and maintenance requirements of equipment hired internally may have to take second place to that required on the open market.

Internal Plant Subsidiary

The dilemma of servicing two different types of client presented by a controlled plant policy is eliminated when the activities of a plant subsidiary are restricted to internal hire only.

This system often results in plant hire rates which bear little relationship to market rates, as the type of equipment items and utilisation levels are dictated by the demands of the construction division. Nominally, the plant subsidiary is required to achieve a set rate of return on the capital employed. Sometimes, however, the targets cannot be achieved and the deficits must be covered by the parent company. Such an arrangement may produce a management team which is not held in the same regard as the profit-oriented parts of the company, with a consequent loss of influence and confidence of the plant manager. Decisions on plant purchases and control may increasingly be imposed by the construction side of the business, to compound the difficulties of making profits from the equipment holdings.

Low Plant Ownership

Some construction companies operate very small equipment holdings, on the grounds that achieving profitability from plant ownership is relatively less rewarding than other construction activities. A small depot may be maintained to provide small items only and most of the problems of owning equipment are avoided. This system, of course, relies heavily on the facilities provided by plant hire companies. However, the availability of specialist equipment items can influence the work load and contract type open to the company, and so could affect the success of this sort of policy.

Rehiring Plant Company

In order to reduce the administrative duplication of each site or contract, obtaining and then invoicing for payment the plant requirements, some companies operate on a basis similar to the low plant ownership arrangements, but provide a centralised service of hiring-in all equipment and passing it on to the sites. The main advantage lies in the ability of a centralised administration to negotiate favourable terms and discounts with regular hire firms. Besides economy, some co-ordination of plant hire requirements across the company can also be achieved and so accommodate the transfer of equipment items from one site to another. An extension of this system is the recent emergence of rehiring companies supplying equipment to the hire sector itself.

No Plant Structure

The parent company could take the view that the policy of using an unstructured organisation with respect to plant holding will serve the firm's needs best. Several arrangements are possible. For example, individual contracts may purchase equipment and be credited subsequently with nominal resale values when the plant leaves the site. In this case care has to be exercised in assessing equitable sums when purchases and resales are internal transactions. This method is usually confined to special items, such as grouting pumps, cableways, etc., which are usually sold off when a contract has no further use for them. In conjunction with this system, more general items may be moved from site to site without a formal charging procedure. Plant is costed as an overhead to the contract on an arbitrary basis, but these policies clearly carry the risk of not forcing the plant to make a sound financial contribution to the company's activities.

1.3   PLANT STRUCTURES IN PRACTICE

The top fifty construction companies (expressed in turnover terms) operate their equipment holdings in the following ways:

Type of plant organisation	Number of firms
(a) Independent plant hirer	13
(b) Controlled plant hirer	10
(c) Internal plant subsidiary	4
(d) Rehiring company	7
(e) Low plant ownership	6
(f) No plant structure	10
	50

It can be seen that almost one half of the firms operate the plant department as a profit centre (a and b). Only a very small proportion (four firms) have a service system of internal plant hire (c). These latter companies are in fact four of the largest firms in the land and presumably have sufficient work for equipment in their own organisations to keep utilisation at levels which are economic and profitable for the firm.

The rehiring and/or low plant options are clearly favoured, especially for small firms or those not requiring much equipment, such as house building firms. These are turning to the plant hire market for their requirements.

The above list of types of plant organisation covers practically the whole spectrum of the industry, but in addition the DIY small tools and equipment sector must be included. Several thousand shops, from small outlets to super stores, have developed in recent years, supplying small items to both individuals and companies for a wide range of duties, from garden equipment to small-scale building equipment. Each method has its own merits and the following questions should be asked before any item of plant is acquired:

(1) Is ownership of that item of equipment fundamental to the operations of the business?
(2) Will the capital locked up in the equipment generate an adequate rate of return compared with other forms of investment?
(3) Is purchasing the equipment for direct ownership the only profitable way of obtaining and operating it?

Unless the reply to these questions is unequivocably positive, some other sound commercial reason should be established before authorisation to acquire the equipment is granted.

BIBLIOGRAPHY

Anthill, S. H., Ryan, P. W. S. and Easton, G. R. (1989). *Civil Engineering Construction*, McGraw-Hill

Barrie, D. S. and Paulson, B. C. (1983). *Professional Construction Management*, McGraw-Hill

Construction Equipment and Methods Journal (monthly)

Construction Management and Economics Journal (quarterly)

Douglas, J. (1978). *Construction Equipment Policy*, McGraw-Hill

Green and Co (stockbrokers) (annual). Investment reports on the construction plant hire industry, London

Harris, F. (1989). *Modern Construction Equipment and Methods*, Longman

Harris, F. and McCaffer, R. (1989). *Modern Construction Management*, 3rd edn, BSP Professional, London

Journal of the Construction Engineering and Management Division (quarterly), ASCE

Mead, H. T. and Mitchell, G. L. (1972). *Plant Hire for Building and Construction*, Newnes-Butterworths

Peurifoy, R. L. and Ledbetter, W. B. (1985). *Construction Planning, Equipment and Methods*, McGraw-Hill

Plant Managers Journal (monthly)

Chapter 2
Hire and Rental of Equipment

2.1 INTRODUCTION

The hiring or rental of construction equipment has developed during the past 40 years and has introduced a new dimension into contracting. A considerable choice is now available in the range of equipment for hire, freeing many small contractors from the burden of having to stock and maintain uneconomic items of equipment.

The real birth of the industry occurred in the 1930s during the economic depression, when many companies had an insufficient workload to justify purchasing all the equipment wanted. This presented an ideal opportunity for entrepreneurs to specialise in holding popular items. During the reconstruction which followed World War II, hired plant was in great demand, and since then the hire market has continued to grow: today an estimated 60% of all plant used is hired.

Independent plant hire firms are a mixture of large and small companies totalling over 3000 firms excluding the DIY small tools sector, as indicated in Table 2.1.

It can be seen from the table that the largest 5% of firms account for slightly more than 50% of the total value of the plant hire business. The size of the turnover represents about 5% of the total value of work undertaken by the construction industry. The high proportion of hired equipment used by British construction companies is not reflected elsewhere, for example, in continental Europe only about 5% of the total plant used is hired. The reasons for the plant hire market developing rapidly are difficult to discover, but the need for a hiring facility would seem obvious. The following reasons have possibly been influential.

Table 2.1   Approximate structure of the plant hire market

Percentage of total firms	No. of employees	Percentage of total turnover	
18	0–1	0.7	
40	2–7	6.6	
14	8–13	9.0	
12	14–24	11.0	
4	25–34	7.4	
6	35–59	13.9	
1.6	60–79	5.9	
1.6	80–114	7.4	51.4
2.2	115–299	20.3	
0.6	300 or more	17.8	
100		100	

(1) Construction companies on the continent tend to take more pride than their British counterparts in displaying the firm's name and livery on equipment used on contracts.

(2) In some countries the law protecting firms hiring out equipment possibly offers few safeguards.

(3) The rate of growth of the UK economy during the past 30 years has been relatively slow compared with other major industrial economies, with the accompanying lower demand for construction facilities. Thus, where economic utilisation levels for many plant items may have been possible, such an advantage has not been available generally in the UK. Indeed, many regions have suffered an erosion of their industrial base, with a consequent loss of economic activity. Such a situation has encouraged the pooling of equipment resources in a hire market.

(4) The UK construction industry includes a large proportion of small firms, which provides a lucrative market for hired equipment.

(5) Government policies in the past allowing full and immediate depreciation of the cost of equipment encouraged purchases to offset tax payments on profits. This was an inducement for firms to buy plant and look for work to maintain profitable levels of utilisation. Such an arrangement would favour the development of a hiring system.

Once a plant market exists, items of equipment cannot readily be disposed of when the economic fortunes of the construction industry deteriorate. Consequently, the supply of plant, ably abetted by government tax policies, has generally grown to meet boom conditions, with a subsequent tightening of hire rates during a recession thereby forcing hire rate adjustments to fall short of the rate of inflation, with the result that firms then have had to extend the life of equipment beyond the original estimates and minimise servicing and maintenance. The net result is that much old equipment has to be hired out at uneconomic rates, the situation finally culminating in amalgamations and bankruptcies. Currently hire rates will vary from region to region, depending upon the state of the market and the supplier. The Construction Plant Hire Association (CPA) and Hire Association of Europe (HAE) publish schedules of hire rates in their handbooks, to offer guidance to both the hirer and supplier, with similar information also readily available in equipment journals.

2.2   THE HIRE OR RENTAL COMPANY

The principal purpose of any hire firm is to supply the equipment needs of construction clients at a profit. The emphasis is therefore market-oriented compared with the 'service' plant division found in a contractor's organisation. The strategy of a single construction company would have relatively little influence, as the plant hire firm should be more concerned to satisfy the demands of the total market. However, the market for construction equipment hire varies in both the opportunities for specialisation and the quality of service demanded. It is the management's responsibility to define the aims and objectives of the firm, so that a suitable organisation may be assembled to operate in a competitive market.

2.2.1 Company Objectives

The decisions to be taken include defining:

(1) The kind of goods and services to offer — for example, earthmoving equipment, craneage, small general plant or specialist equipment, DIY equipment etc. — and the corresponding location and organisation of stockyards, rental outlets and service facilities.
(2) The desired share of the market.
(3) The possible changes in and fluctuations of the market in future years.

Once these are established, the company's long-term plans may be formulated. These will involve the preparation of a market forecast to be matched subsequently with a corresponding corporate analysis to highlight the strengths and weaknesses of the company for coping with the potential market opportunities (see Figure 2.1).

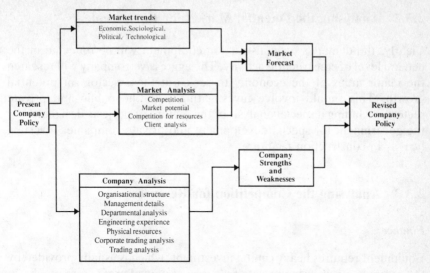

Figure 2.1   The market forecast

2.3   THE MARKET FORECAST

The market forecast should endeavour to seek out the wants and needs of the market for hired equipment and should be a systematic and continuous process, executed religiously, if the firm is to survive and prosper. The task can be taken in separate stages from which information is finally synthesised to produce the new policies and strategies. The main areas for a typical hire company are shown below.

2.3.1   Analysing the Competition for Hire Services

A brief survey of other equipment hire and rental companies may reveal segments of the market which have not been fully exploited or, conversely, should be avoided because of fierce competition. The main points to determine are the strong and weak areas of the hire market, including the following aspects:

(1) The firm's market share of the different lines of plant and equipment held, together with an analysis of the recent performance of each type with respect to growth and profitability.
(2) The market share, growth in turnover and profitability of the major competitors, noting the areas of interest for each.
(3) The margin of differences between the company's hire rates and those of the competition, to give a guide to the improvements required.

2.3.2   Analysing the Potential Market for Equipment

Clearly, the demands for construction equipment will be reflected in the general level of construction activity. The aggressive company will research the major areas of the economy to seek out those sectors of potential growth. This should involve investigating both the public and private sectors, including a special analysis of the regions or projects designated by the government for special development and private companies investing heavily in construction facilities.

2.3.3   Analysing the Competition for Resources

Finance

Equipment requires heavy capital investment, which is usually provided by the company itself, using private resources, retained profit, hire purchase, leasing, etc., or by a bank loan. The availability of loan capital is likely to fluctuate according to the fortunes of the national economy, with unpredictable changes in the interest rate, and lenders may prefer other sectors of the economy to the construction industry. Furthermore, only companies with a sound financial record of profitability, with mortgageable assets, are likely to be favourably considered by the banking and financing sector. Also, before any item is acquired, it should be remembered that a machine once purchased often cannot be turned quickly into liquid cash assets to deal with a crisis.

Plant, Personnel and Premises

Few companies possess the resources or expertise to operate in all sectors of the plant hire or equipment rental markets. Therefore, external factors which would affect the firm's ability to compete must be defined. For example, the reliability of the various manufacturers and supply agents should be assessed, since the quality of back-up services and availability of spares will have a considerable bearing on competitive performance.

Construction equipment will usually last longer and be less costly to maintain if the machine operators and servicing staff are well-trained and responsible. There is always competition for such skilled personnel, and a company not prepared to train, educate and pay its workforce well should avoid sophisticated and technically complex plant.

New premises may be required to establish a new plant company, rental shop or area division. Often the depot must be located near the main market, such as a large town. However, sites which can provide room for expansion, good access and security will be in demand from other firms and industries. These aspects are often overlooked when expansion programmes are put into operation.

2.3.4   Analysing the Client

Some clients are better 'payers' than others, which may help to reduce the need for cash or overdrafts maintained by the plant firm. Although there may be an apparently lucrative market, such as a national road programme, for certain equipment lines, clients may be so slow and awkward in their attitude to payment that the plant firm would be advised to avoid them.

2.4   STRENGTHS AND WEAKNESSES OF THE COMPANY

It is essential to consider the ability of the organisation to cope with a change of direction or expansion. The structure of the firm must be examined for both its overall structure and the strength of each department.

2.4.1   The Corporate Analysis

Organisation Structure

Most companies have a family tree which represents the official structure of the management organisation. In practice, the actual lines of command and communication are likely to be more subtle than those formally recognised. However, this family tree is a good starting point in highlighting potentially weak structural arrangements.

Management Details

The quality of present managers will be tested when entering new markets. Much information is often held by the personnel department on such matters as salary, qualifications, education, training and experience. These data help to identify potentially strong management areas and those which have failed to develop a healthy ladder of achievement on which the younger men can gain experience. If the process is repeated for each department, gaps and stagnant areas become apparent.

Financial and Operational Control Departments

This review should be extensive and probably should not be undertaken until policies have been made tentatively. The most likely candidates for investigation are the accounts, administration and equipment servicing departments, since they tend to be labour-intensive and reluctant to accept rapid changes. Such departments contribute largely to the overheads, which may rapidly increase if the company expands into new markets. Overheads should also be borne in mind when moving from a fairly low technical market — say small machines — into a specialised market requiring high technical competence and support.

Engineering Experience

Management and operational control surveys may yield much information about the nature of the company and its employees. The plant business requires that good managers should also be good engineers. Any change in policy should spring from a sound base of experience: it is far too risky to rely entirely on imported skills when undergoing change. Therefore, a careful analysis is required of the existing skills within the company to see whether they will provide an adequate basis on which to build. In particular, staff expertise is likely to be severely tested when policy changes involve the introduction of new equipment lines or when the firm decides to decentralise and establish depots sited away from headquarters.

Physical Resources

Putting new objectives into practice may necessitate new depots, outlets and storage facilities. However, the acquisition of land and the construction of new facilities take time, are expensive and demand careful planning of the location. In addition, new and different equipment items may require new maintenance facilities, which may be costly and beyond the knowledge and experience of the management and workforce.

Corporate Trading Analysis

The following financial ratios yield important information in assessing the financial strength of the company and comparing its performance with that of major competitors:

- Return of capital employed.
- Profit on turnover.

- Turnover of capital.
- Growth in capital employed and in net profits.
- Current assets to current liabilities.
- Stock values to sales ⎤
- Debtors to sales ⎬ converted to time periods.
- Profit per employee ⎦

By comparing figures over the past 5 years with other companies in similar fields, some judgement is possible on the viability of the firm and its ability to take on new ventures successfully.

Trading Analysis

The trading analysis means looking at individual equipment lines in a fair degree of detail. The types of questions to be asked are:

- What trends in profitability — say during the past 5 years — can be seen in the various equipment types and lines?
- How did actual profit compare with estimates?
- How has inflation affected costs and what was the policy towards hire revenues?
- What effect would changing the mark-up included in plant hire and equipment rental rates have had on turnover and overall company profits?
- How did maintenance costs, actual machine life and utilisation levels compare with estimates?

2.5   TRENDS AFFECTING THE FORECAST

The stage of proposing any changes in company policy, as shown in Figure 2.1, has now been reached. Once the facts are known, experienced managers will usually see what changes need to be made. When these tentative proposals have been put forward, managers should realise that new facts will emerge and errors in the forecasts will appear. These are inevitably caused by political changes, shifts in the market outlook, technological developments and economic influences. The effects of these movements are difficult to quantify, but should be kept under cautious review, the policies being adjusted where necessary. Care should be taken not to overreact to new events, as this can cause loss of confidence at middle management level.

2.6   PROMOTING THE COMPANY'S SERVICES AND SATISFYING THE CUSTOMER

Once the company has established the plant and equipment lines and the market it desires to service, it becomes vitally important to increase the awareness of the potential customer. This may be achieved by a variety of advertising methods coupled with fostering good public relations. The latter will probably only bring results in the medium-to-long term and should include providing clean, reliable and well-maintained equipment with an efficient back-up service of spare parts and advice. Co-operation with the client or customer is always helpful. Most construction contracts often involve slight delays and changes to the original requests for plant operation and hire. These are not always detrimental to the operating costs of the plant firm, and the goodwill generated will help in the future.

Many customers and clients are impressed by 'added-values', and clearly the firm with back-up services of experienced and well-qualified staff in servicing, maintenance, law, insurance, technical advice, etc., will be a more credible company than one without such facilities.

BIBLIOGRAPHY

Cox, V. C. (1983). *International Construction*, Longman

Fisher, N. (1986). *Marketing for the Construction Industry*, Longman

Institute of Marketing (1974). *Marketing in the Construction Industry*

Jepson, W. B. and Nicholson, M. P. (1983). *Marketing and Building Management*, Mechanical and Technical Publishing Co

Moore, A. B. (1984). *Marketing Management in Construction*, Butterworth

Smallbone, D. W. (1983). *An Introduction to Marketing*, Staples

Thomas, M. (1986). *Pocket Guide to Marketing*, Blackwell and Economist

Urry, S. and Sherratt, A. T. C. (1980). *International Construction*, Construction Press

Chapter 3
Organisation of Hire Companies and Departments

3.1 MANAGEMENT STRUCTURE

The appropriate management structure will depend upon the nature and size of a firm's business activities. In particular, an independent rental or hire firm will require all the management functions of a market-oriented company, as shown in Figure 3.1. The internal plant division of a construction company, however, is generally integrated into the parent company's activities, and functions such as purchasing and financial accounting may be the responsibility of the parent firm. For both types of business, the need to decentralise into geographical regions, or even major equipment categories, is a further complication, and firms tend to make individual depots or sites responsible for their own business activities when

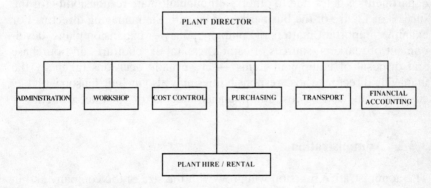

Figure 3.1   Management structure of a plant hire company

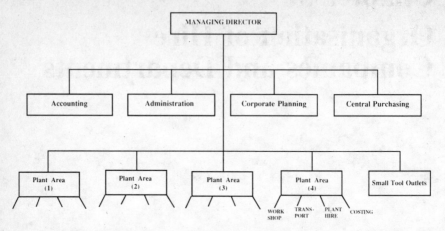

Figure 3.2 Management structure of a regionalised plant holding

faced with this situation, although responsibility for overall company policy, major purchasing and financial accounts may remain at head office (see Figure 3.2).

3.2   TYPICAL MANAGEMENT FUNCTIONS AND DEPARTMENTS

3.2.1   The Managing Director

The managing director sets the objectives of the business and assures that necessary strategies are adopted to ensure that the firm will survive and prosper. The managerial functions involved in all but the smallest firms require that much of the day-to-day responsibility for running the various departments is delegated to others, although ultimate responsibility for the success or failure of the business must lie with the managing director. For example, important matters of company policy — expansion plans, development of markets, sources of capital, capital expenditure, and purchase and disposal of equipment items — are usually decisions taken by the managing director, whose responsibilities may also include liaison with key customers and setting cost and financial budgets for each department.

3.2.2   Administration

The administrative function will grow with the size of the company and in large firms it is subdivided into separate elements. The list of duties includes health and welfare of personnel, safety regulations, employee training, provision of social facilities, a postal service, legal and insurance advice, the negotiation of wages and salaries, conditions of employment, personnel record keeping and maintenance of the company's physical assets. Authority is vested in a personnel/administration manager with subordinate managers who are responsible for these duties.

3.2.3   Financial Accounting

The company accountant is responsible for the payment of invoices, receipts from hire sales, control of cash and bank overdrafts, and preparation of the trading, profit and loss accounts and balance sheet in accordance with the Companies Acts. The accountant has an important function and often works closely with the managing director in controlling the overall financial affairs of the company. For example, he will be involved in making decisions on the type and source of capital for major purchases, the financial viability of an expansion programme and preparing the company financial budget for the year ahead.

3.2.4   Purchasing

The buying department is responsible for obtaining quotations for materials, supplies and other consumables used at the depot and on plant located in the field but maintained and supplied from the central or regional depot. Thus advantages of centralised purchasing may be realised from: (1) The ability to obtain discounts from bulk purchasing; (2) The efficiency generated by adopting standard procedures; (3) Monitoring the quality of supplies; (4) Experience gained from the commercial operations of suppliers, and, of course, (5) Centralised administration facilities. However, when a plant department is relatively small and items on site are maintained under the responsibility of the construction department, a separate purchasing department is hardly necessary and consumables are purchased as required by site and charged to the workshop maintenance account.

A secondary function of the purchasing department may be participating in the purchase and sale of equipment, especially inviting quotations and assessing the commercial details of a transaction, although, the advice of the heads of other departments will also be involved at this level of plant procurement and disposal.

3.2.5 Cost Accounting

The cost accounts department collects and interprets data from the other departments, and its tasks include the preparation of targets in the form of budgets against which costs and revenues may be monitored. This information is used by departmental managers to control and update their operations so that cost targets may be achieved. In particular, the department must record all the data required to prepare hire rates for equipment, and returns are therefore required from all departments, including the hours operated weekly for each item.

3.2.6 Workshop Control

The workshop manager is primarily responsible for the maintenance and servicing of the firm's equipment. He must provide and maintain a store of consumable materials and spare parts, with appropriate stock control procedures, although the actual purchasing responsibility usually lies with the buying department. Costs incurred by the workshop include the wages of fitters, mechanics, electricians and other operatives needed to perform servicing and repair duties, plus the costs of tools, materials, mobile workshops, general overheads, salaries paid to staff and foreman, and equipment inspection and administration. The budget for the department is allocated from past records of the equipment holdings and any additions that can be foreseen. It is, therefore, essential that maintenance records of each item be rigorously updated to facilitate the monitoring of costs against the budget. As a rough guide, maintenance facilities for about 5% of the fleet should be provided.

The workshop manager's function may be augmented by a field manager to provide advice to construction sites on the use, operation and routine maintenance of plant and equipment. In particular the duties may complement those of the workshop manager, where the company operates a comprehensive system of mobile maintenance to sites. A complaints service may also be provided by the maintenance department.

Table 3.1 Example of a company asset register

Machine number	Machine description	Purchase date	Purchase price (£)	Scrap/ resale value (£)	Machine life	Type	Depreciation charge (£)	To date (£)
601	Terex TS14/70	09 1974	100 000	20 000	5	DBAL	16 000	60 000
602	CAT 633C Scraper	06 1974	120 000	20 000	5	DBAL	20 000	55 000
603	Terex TS14/75	11 1975	118 000	18 000	5	DBAL	20 000	40 000
604	Terex IS14	11 1975	118 000	18 000	5	DBAL	20 000	40 000

3.2.7   Transport

Most construction equipment is unsuitable for travel on public roads and must be transported from site to site on trucks and lorries. The transport supervisor works with the workshop and plant hire departments to co-ordinate the transport needs of the construction sites and other clients. The costs allocated to this department include the running costs of the transport fleet, such as fuel, maintenance, servicing, drivers' wages, supervisors and administration staff salaries, capital cost of the transport fleet and overheads. The responsibility for servicing and maintaining the plant fleet will generally remain with the transport department, but execution of the work may be undertaken by the works department and subsequently charged to the transport account.

3.2.8   Hire

The plant hire department or equipment rental shop provides the selling function and is responsible for generating sufficient revenue from hire of items to earn a profit for the company. In particular, the hire manager must work closely with the managing director and cost control department when operating in the open hire market, as it is vital to set flexible and competitive rates. The hire manager should be provided with an asset register of records on equipment availability and location, together with information on utilisation levels, cost returns and maintenance needs. The costs allocated to this department are those generated by the salaries of the hire staff, marketing, advertising and overheads.

3.2.9   The Asset Register

The efforts of the various departments and functions in the enterprise described above should be directly or indirectly concerned with the profitability, utilisation and performance of the firm's equipment. In order

Written down value (£)	Machine hire rate (£)	Budgeted earnings to date (£)	Actual earnings to date this month (£)	Total costs to date (£)	% P/L on earnings	% Utilisation Year	Month	Location
4 000	45.36 h	4 600	5 000	4 500	10	82	85	Bristol
45 000	47.49 h	4 000	3 950	3 500	11.4	75	76	Birmingham
60 000	48.33 h	5 000	6 000	5 900	1.7	76	73	Wol-verhampton
60 000	48.33 h	4 900	4 000	4 100	− 2.5	70	67	Lough-borough

to enable the desired procedures for controlling these functions to be co-ordinated, most firms prefer to record key data on each item on an asset register, as typified by Table 3.1. The recording of the information may be manual, but it is becoming advantageous to store data on computer files. The basic information required on the asset register should include, for each item, a code number, a registration number, make, model and short description. However, because at some point most departments in the firm will need to refer to the register, additional recorded data are necessary and separate reports must be added. For example, the financial accountant requires data on purchase date and price, planned life, depreciation method, book value, depreciation charge and depreciation to date. The hire department requires the current utilisation factor, hire rate, budgeted earnings, actual earnings and actual costs. In addition, an inflation index for the particular equipment category is useful in setting new hire rates in times of inflation. The workshop needs current data on location, base depot, planned operating hours and actual operating hours so that effective maintenance can be monitored. An up-to-date asset register can act as a sound programme for carrying out safety inspections required for insurance and by law. And, not least, the managing director must be constantly aware of an item's profitability and utilisation level.

BIBLIOGRAPHY

Drucker, P. T. (1989). *The Practice of Management*, Heinemann
Gerloff, E. A. (1986). *Organisational Theory and Design*, McGraw-Hill
Hardy, C. (1981). *Understanding Organisations*, Penguin
Hicks, H. and Gullett, C. (1981). *Management of Organisations*, McGraw-Hill
Kootz, H. and O'Donnell, C. (1986). *Essentials of Management*, McGraw-Hill
Robbins, S. P. (1987). *Organisational Behaviour*, Prentice-Hall
Sofer, C. (1973). *Organisations in Theory and Practice*, Heinemann

Section 2

Investment and Procurement

Chapter 4
Economic Comparisons of Equipment Alternatives

4.1 PRINCIPLES OF ECONOMIC COMPARISONS

The basic approach to economic comparisons is to assemble all the costs relating to one course of action and all the costs relating to the alternative course of action and to compare them. The assembling of the costs must be in such a way that the two are comparable. It is the difficulty of ensuring that the assembled 'packages' of costs are comparable that requires the calculation of either present worth, or value, of two proposals or the equivalent annual cost of the two proposals. Of these techniques of comparison, present worth is more commonly used. Both present worth and equivalent annual costs require an interest rate which is taken to represent the value of money to the investor. That is, it represents the interest the investor could receive elsewhere.

Comparisons not involving interest rates are very common in short-term schemes – that is, schemes of less than 1 year. Construction site staff are continually and almost subconsciously undertaking economic comparisons without interest in the calculations. Such economic comparisons include comparing the hire rate for different cranes, or the hire rate for different excavators, or the hire rate for an excavator from an external equipment hire company with the hire rate from the company's own equipment division. Comparing the cost of, say, hand excavation with the cost of using an excavator is another economic comparison. All these comparisons refer to operations with relatively short time periods of a few weeks or a few months and the effect of interest is not significant. Thus, the comparisons are valid and easy to make on an equitable basis. The comparisons become more difficult when the operations or schemes to be compared last a few years or more, when the effect of interest becomes significant and needs to be included in the calculation. The difficulty is assembling the various costs into 'packages' that can be compared for these longer-duration operations that require 'present worth' or 'equivalent annual costs'. Some examples of longer-duration operations or schemes are equipment required for quarrying, open-cast mining or concrete production.

All the examples in this section are based on cash flows that have been estimated at present or year zero prices without taking inflation into account. The interest and time relationships used in this chapter are explained in the Appendix, which includes examples of interest tables for interest rates of 10% and 15%.

4.1.1   Present Worth

Present worth comparisons are used to compare two or more schemes where the equipment chosen for each scheme leads to different capital investment and different running costs. Essentially, present worth enables the trade-off between capital investment and future running costs to be compared. Example 4.1, which includes the capital cost of buying equipment and the running costs of operating the equipment, illustrates the comparison. All the estimates used in this comparison are at present-day (i.e. year zero) prices.

EXAMPLE 4.1

Assuming an interest rate of 15%:

	Proposal 1	Proposal 2
Capital cost of equipment	£8 500	£9 500
Annual running costs	£1 750	£1 500
Life	5 years	5 years

The present worth of Proposal 1 is £14 366 and of Proposal 2 is £14 528. The calculations are as follows:

Proposal 1

$$\text{Present worth} = £8\,500 + (£1\,750 \times 3.352)$$
$$= £8\,500 + £5\,866$$
$$= £14\,366$$

Proposal 2

$$\text{Present worth} = £9\,500 + (£1\,500 \times 3.352)$$
$$= £9\,500 + £5\,028$$
$$= £14\,528$$

The capital sums are already in year zero and need no further manipulation. The running costs of £1 500 and £1 750 each year need to be converted to present worth or capital sums. The factor used for this conversion is the uniform series present worth factor. This factor, 3.352, has been taken from the tables in the Appendix.

The present worth of Proposal 1 is £14 366 and the present worth of Proposal 2 is £14 528. Thus, Proposal 1 is the more economic. The present worth of Proposal 1, £14 366, represents enough money to buy the equipment item at a cost of £8 500 and investing the remainder at 15% is enough to produce £1 750 each year for the next 5 years. Thus £14 366 is the amount required now to meet all the requirements of Proposal 1. Similarly, the present worth of Proposal 2, £14 528, is the amount required now to meet all the requirements of Proposal 2. Since the present worth of Proposal 1 is the smaller and both proposals would be compared only if the two items of equipment were capable of doing the same tasks, then the one with the least cost, i.e. least present worth, is the most economic.

An alternative way of considering this comparison is to examine the differences between the capital costs and the running costs. Proposal 1 has £1 000 less capital but requires £250 more running costs each year. Thus, the extra capital of £1 000 involved in Proposal 2 can be seen to be buying £250 of savings in the running costs. The question as to which of the two schemes is the more economic could be restated as follows: would the £1 000 of extra investment be better used in saving £250 in running costs or earning 15% if invested elsewhere? £1 000 invested at 15% for 5 years would give an income of £298.31, calculated as

$$£1\ 000 \times 0.298\ 31 = £298.31$$

where 0.298 31 is the capital recovery factor, taken from the tables in the Appendix. Thus, the return on the investment is better than the saving in the running costs. Therefore, Proposal 1, the smaller of the capital investments, is the more economic proposal.

This present worth comparison is valid so long as the lives of the two proposals are the same. This is usually the case in comparing equipment items. One exception is comparing the cost of keeping an item of equipment for 1 year with that of keeping it for 2, 3 or 4 years. In such cases the lives are different and require different treatment. This is explained later, in the section dealing with replacement.

The example presented is the use of present worth in a simple case where there are only capital and running costs and where the running costs were uniform — that is, the same each year. The principles of using present worth are the same even when the cash flow becomes more complicated. The next example shows the running costs varying each year in order to reflect the increasing costs incurred as the equipment grows older. Also included in the next example is a resale value of £4 000 occurring in the last year. The resale value is a return of money to the investor and therefore carries a different sign from the capital and running costs, which are outflows of money. Again all the cash flows are estimated at present prices.

EXAMPLE 4.2

Assuming an interest rate of 15%:

Cash flow for the purchase and
resale of an item of equipment

Year	(£)
0	− 8 500
1	− 1 750
2	− 1 850
3	− 2 000
4	− 2 200
5	− 2 500 + 4 000

The present worth of these cash flows is £13 247.72, calculated as shown below at an interest rate of 15%. This present worth can now be compared with the present worth for an alternative proposal:

Year	Cash flow (£)		Present worth factor (15%)		Present worth (£)
0	− 8 500	×	1.0	=	− 8 500.00
1	− 1 750	×	0.86956	=	− 1 521.73
2	− 1 850	×	0.75614	=	− 1 398.87
3	− 2 000	×	0.65751	=	− 1 315.02
4	− 2 200	×	0.57175	=	− 1 257.85
5	− 2 500	×	0.49717	=	− 1 242.93
5	+ 4 000	×	0.49717	=	+ 1 988.68
			Total present worth	=	−£13 247.72

The present worth factors are taken from the tables in the Appendix. Because the annual sums are varying in this case, the uniform series present worth factor cannot be used and the individual present worth factor for a lump sum, $1/(1 + i)^n$, is used instead, where i is the interest rate and n is the number of years. This increases the arithmetic involved but is unavoidable when dealing with varying annual cash flows.

The treatment of the £4 000 resale value shown here is to simply add it into the present worth, taking account of the different sign. It may be more acceptable to deduct its present worth for the initial capital:

Initial capital invested   = £8 500.00
Present worth of resale = £4 000 × 0.49717 = £1 988.68
Adjusted capital invested   = £6 511.32

The total present worth calculated after adjusting the capital invested in this way to take account of the resale value will be the same as the example given.

4.1.2   Equivalent Annual Costs

An alternative to present worth comparison is comparing proposals on the basis of the equivalent annual costs. Whereas present worth converts all future running costs to a present worth or capital sum, equivalent annual costs convert the capital sums to an annual cost. Equivalent annual cost comparisons achieve the same as present worth comparisons and, like present worth comparisons, are essentially evaluating the trade-off between capital and running costs. Example 4.3 uses the same cash flows estimated at present prices as Example 4.1.

EXAMPLE 4.3

Assuming an interest rate of 15%:

	Proposal 1	Proposal 2
Capital cost of equipment	£8 500	£9 500
Annual running costs	£1 750	£1 500
Life	5 years	5 years

The equivalent annual cost of Proposal 1 is £4 285.64 and the equivalent annual cost of Proposal 2 is £4 333.95. The calculations are as follows:

Equivalent annual cost
of Proposal 1   = £1 750 + (£8 500 × 0.298 31)
                         = £1 750 + £2 535.64
                         = £4 285.64

Equivalent annual cost
of Proposal 2 $= £1\ 500 + (£9\ 500 \times 0.298\ 31)$
                       $= £1\ 500 + £2\ 833.95$
                       $= £4\ 333.95$

The annual running costs of £1 750 and £1 500 do not need further manipulation. The capital costs of £8 500 and £9 500 need to be converted to annual costs. The factor used for this conversion is the capital recovery factor, 0.298 31, and has been taken from the tables in the Appendix.

The equivalent annual cost (EAC) of Proposal 1 is £4 285.64 and the EAC of Proposal 2 is £4 333.95. Thus, Proposal 1 is the more economic, as was found by the present worth comparison. The EAC represents the annual cost of owning and operating the item of equipment. This is made up of £1 750, representing the annual running cost, and £2 535.64, representing the annual cost of the capital investment. This is calculated on the basis that if £8 500 were invested at an interest rate of 15%, an income of £2 535.64 could be taken each year for the next 5 years. This income would be made up of the interest earned plus the original capital. At the end of the first year £1 275 of interest would be earned on the £8 500 of capital invested. Thus, if an income of £2 535.64 were taken, this would be made up of £1 275 of interest plus £1 260.64 of capital, leaving £7 239.36 of capital. At the end of the second year the £7 239.36 of capital would earn £1 085.90 of interest. The income of £2 535.64 would be made up of the £1 085.90 of interest plus £1 449.74 of capital, leaving £5 789.62. The interest earned in the third year would be £868.44 and the income of £2 535.64 would be made up of the £868.44 of interest plus £1 667.19 of capital, leaving £4 122.43 capital. In the fourth year the interest earned would be £618.36 and the income of £2 535.64 would be made up of the £618.36 interest plus £1 917.27 of capital, leaving £2 205.16. In the fifth year the interest earned would be £330.77 and the income would be made up of the £330.77 interest plus £2 204.86 capital, leaving the account exhausted and no capital (on the basis of the calculations presented here, £0.30 of capital would remain but this is due simply to rounding errors in the calculation). Thus, if the £8 500 were invested at 15%, an income of £2 535.64 could be taken each year for 5 years. However, since the £8 500 was not invested in such an account but used to purchase the item of equipment, the investor is deprived of the income of £2 535.64 and therefore this can be regarded as equivalent to the annual cost of owning the equipment item.

This example shows the use of equivalent annual cost where only capital and running costs are considered and the running costs are uniform. The use of equivalent annual costs becomes more difficult when running costs vary from year to year. In such cases it is necessary to convert the varying running costs to a capital cost before converting them back to an EAC. The next example, using the same cash flows as Example 4.2, illustrates this difficulty. As in Example 4.2, the cash flows are estimated by use of present prices.

EXAMPLE 4.4

Given that the interest rate is 15%:

Year	Cash flows for the purchase, operating and disposing of an item of equipment (£)
0	− 8 500
1	− 1 750
2	− 1 850
3	− 2 000
4	− 2 200
5	− 2 500 + 4 000

The present worth of the resale is £1 988.68 and the purchase price less resale, i.e. the adjusted capital invested, is £6 511.32, as explained in Example 4.2.

The EAC of the capital invested is £1 942.39, calculated as follows, taking the capital recovery factor for 5 years, 0.298 31, from tables in the Appendix:

$$EAC = £6\ 511.32 \times 0.298\ 31 = £1\ 942.39$$

The present worth of the running costs are £6 736.40, summed from the calculations in Example 4.2 and the equivalent annual cost of the present worth of these running costs is £2 009.54, calculated as follows, taking the capital recovery factor for 5 years, 0.298 31, from tables in the Appendix:

$$EAC = £6\ 736.40 \times 0.29831 = £2\ 009.54$$

Thus, the total equivalent annual cost for all the cash flows — the purchase price less resale value and running costs — is £3 951.93.

This is made up of £2 009.54, representing the running costs, and £1 942.39, representing the purchase price less resale. Thus, the EAC of £3 951.93 can be used for comparison with similarly calculated EACs for alternative proposals.

By converting the varying annual running costs to a present worth and then converting this present worth to an EAC, the running costs can be distributed uniformly over each year. This procedure overcomes the difficulty with varying running costs. However, this procedure also illustrates why the use of present worth as a basis for comparison is more common. To arrive at an EAC, the cash flows had first to be converted to a present worth. Therefore, it is easier to perform the comparison on the basis of present worth rather than involve the extra calculation of producing EACs.

However, the EAC method does not require the lives of proposals under comparison to be equal, as is required in a present worth comparison. The reason is that the sum calculated, the EAC, refers to one year and the EAC of any alternative scheme also refers to one year. The process of calculating EACs takes account of the duration of the equipment item and produces annual costs which can be compared with annual costs for other proposals. An example of this is comparing the cost of keeping an item of equipment for 1, 2, 3, 4 or 5 years, thereby determining the best replacement age. The costs of keeping an item of equipment for different periods are, in effect, different proposals which have different lives and therefore present difficulties in comparison on the basis of present worth.

Another use of EACs is to convert the purchase price and running costs to annual costs that can then be compared with the cost of hiring. The same device of converting the capital costs to annual costs is one way of determining the capital element in a hire rate when determining what the economic hire rate should be.

4.2   ECONOMIC COMPARISONS AND INFLATION

In the previous section all the examples were based on cash flows that were estimated at present or year zero prices. The effect of inflation was not taken into account. In Examples 4.2 and 4.4 the annual costs increased as a result of incurring increasing running costs as the equipment grew older and not the effect of inflation. A simple illustration of inflation is that, if in year zero a 'basket of goods' cost £100 and in year 1 the same basket of goods cost £110, then inflation of 10% has occurred. In other words, inflation causes more money to be paid out for the same goods. As individual items change price, or inflate, at different rates, the basket of goods concept is used to create indicators of the average price movements or inflation. The most commonly known 'basket of goods' is that used to calculate the retail price index. Other indices of inflation used in the construction industry are the 'cost of new construction index', the 'tender price index' or the NEDO indices used to calculate price adjustments in construction contracts. All these indices refer to different goods or 'baskets of goods', and indicate the variable nature of inflation between different goods over the same time periods. In the UK the years since 1973, in particular, have also indicated how volatile inflation can be and how elusively difficult it is to predict. This difficulty in prediction cannot be overcome by adjustments in the calculations supporting economic comparisons but, given assumptions as to the forecast inflation, these assumptions can be incorporated into the comparisons.

The approach of estimating cash flows for purchase price, running costs and resale at present or year zero figures and then including the adjustments for inflation is frequently used, because it separates the difficulties of estimating the cash flows resulting from the selection of the equipment item and its technical capabilities from the vagaries of inflation. This separation also allows different inflation assumptions to be made and evaluated without disturbing the underlying estimates.

The following methods explain the means by which the inflation assumptions can be incorporated into economic comparisons. The explanations are based on the cash flows for Examples 4.1 and 4.2.

4.2.1   Method 1: Ignore Inflation

Example 4.1 compared the cash flows from two proposals to purchase and operate a piece of equipment for 5 years. The comparison was made on the basis of present worth. The present worth of Proposal 1 was £14 366 and the present worth of Proposal 2 was £14 528. The cash flows leading to these present worths were based on estimates at year zero prices, the interest rate was taken as 15%, and this comparison, which excluded inflation, indicated that Proposal 1 was the most economic.

Inflation would increase the running costs of both proposals and so increase the present worths. It should be remembered that the present worth is the sum of money required to be invested now to generate the stated cash flows, given a certain interest rate. Thus, if the cash flows are increased, the amount required for investment to generate the cash flows must also increase, given that the interest rate remains the same.

If the purpose of the exercise is simply to *compare* the proposals and select the most economic, and the inflation rate assumed is small, the comparison will not be affected seriously by the inclusion of inflation. That is, if *small* inflation allowances are added into both proposals, the difference between them will not be affected enough to change the comparison. This may be sufficient for small inflation rates but is unlikely to be satisfactory for larger inflation rates experienced from 1973 to the present.

4.2.2 Method 2: Adjusting the Cash Flows

EXAMPLE 4.5

The cash flows in Example 4.1 were:

Year	Proposal 1 cash flow (£)	Proposal 2 cash flow (£)
0	8 500	9 500
1	1 750	1 500
2	1 750	1 500
3	1 750	1 500
4	1 750	1 500
5	1 750	1 500

These cash flows were all estimated at year zero prices. Any cash flow not occurring in year zero would be subject to inflation.

If an inflation rate of 10% per annum were assumed the cash flows would be adjusted as follows:

Year	Proposal 1: Original cash flow + inflation adjustment (£)	Proposal 2: Original cash flow + inflation adjustment (£)
0	8 500	9 500
1	1 750 + 10%	1 500 + 10%
2	(1 750 + 10%) + 10%	(1 500 + 10%) + 10%
3	((1 750 + 10%) + 10%) + 10%	((1 500 + 10%) + 10%) + 10%
4	(((1 750 + 10%) + 10%) + 10%) + 10%	(((1 500 + 10%) + 10%) + 10%) + 10%
5	((((1 750 + 10%) + 10%) + 10%) + 10%) + 10%	((((1 500 + 10%) + 10%) + 10%) + 10%) + 10%

When calculated, these figures become:

	Proposal 1			Proposal 2		
Year	Original cash flow (£)		Inflation adjustment (£)	Adjusted cash flow (£)	Original cash flow (£)	Inflation adjustment (£)	Adjusted cash flow (£)
0	8 500	+	0	= 8 500.00	9 500	+	0	= 9 500.00
1	1 750	+	175.00	= 1 925.00	1 500	+	150.00	= 1 650.00
2	1 750	+	367.50	= 2 117.50	1 500	+	315.00	= 1 815.00
3	1 750	+	579.25	= 2 329.25	1 500	+	496.50	= 1 996.50
4	1 750	+	812.18	= 2 562.18	1 500	+	696.15	= 2 196.15
5	1 750	+	1 068.39	= 2 818.39	1 500	+	915.77	= 2 415.77

Now that the cash flows have been adjusted for inflation, the present worth of both proposals can be calculated as before. Because the cash flows vary from year to year and are not uniform, the present worth factors for each year will have to be used, as in Example 4.2.

The present worths of both proposals are calculated as follows, using an interest rate of 15%:

Proposal 1

Year	Cash flows (£)		Present worth factors (15%)		Present worth (£)
0	8 500.00	×	1.0	=	8 500.00
1	1 925.00	×	0.86956	=	1 673.90
2	2 117.50	×	0.75614	=	1 601.13
3	2 329.25	×	0.65751	=	1 531.51
4	2 562.18	×	0.57175	=	1 464.93
5	2 818.39	×	0.49717	=	1 401.22
			Total present worth	=	£16 172.69

The present worth factors were taken from the tables in the Appendix.

Proposal 2

The present worth for Proposal 2, similarly calculated, is £16 076.58.

This comparison reverses the choice indicated in Example 4.1 by indicating that at 10% inflation Proposal 2 becomes the more economic. Proposal 2, having the smaller running costs, suffers less from inflation than Proposal 1. At smaller inflation rates the choice would still be Proposal 1, as before. However, with an inflation rate at 10% Proposal 2 becomes the more economic. Originally, the present worth of £14 366 for Proposal 1 was sufficient to provide for the £8 500 purchase price and the £1 750 running costs each year, and the present worth of £14 528 for Proposal 2 was sufficient to provide for the £9 500 purchase price and the £1 500 running costs each year. However, 10% inflation causes the present worth of Proposal 1 to become £16 172.69 and provides for the £8 500 purchase price, the £1 750 running costs each year and the additional running costs incurred due to inflation. The £16 076.58 present worth of Proposal 2 provides the £9 500 purchase price, the £1 500 running costs each year and smaller additional running costs due to inflation.

4.2.3   Method 3: Adjusting the Interest Rate

The previous comparison including inflation could have been achieved by adjusting the interest rate rather than the cash flow.

EXAMPLE 4.6

In Example 4.5 the cash flows from Example 4.1 were adjusted to include an inflation rate of 10% per annum, as follows:

	Proposal 1	
Year	Original cash flows (£)	Adjusted cash flows (£)
0	8 500	8 500.00
1	1 750	1 925.00
2	1 750	2 117.50
3	1 750	2 329.25
4	1 750	2 562.18
5	1 750	2 818.39

This adjustment was achieved by adding 10% to the first year cash flows and 10% + 10% to the second year cash flows, and so on. This can be represented as follows, where d represents the inflation rate (0.1 for 10%):

	Proposal 1	
Year	Original cash flow (£)	Inflation adjustment
0	8 500	
1	1 750	$\times$ $(1 + d)^1$
2	1 750	$\times$ $(1 + d)^2$
3	1 750	$\times$ $(1 + d)^3$
4	1 750	$\times$ $(1 + d)^4$
5	1 750	$\times$ $(1 + d)^5$

To calculate the present worth of each of these adjusted cash flows, multiply each year by the present worth factor, as, for example, year 3:

Year	Cash flow	Present worth factor
3	£1 750 $\times (1 + d)^3$ $\times$	0.657 51

The present worth factor, 0.657 51, was taken from the tables in the Appendix or calculated from the expression $1/(1 + i)^n$, where i is the interest rate and n is the number of years. In this case $i = 0.15$ (for 15%) and $n = 3$.

The present worth for year 3 can be calculated thus:

Year	Cash flow		Present worth factor
3	£1 750.00 $\times (1 + d)^3$	$\times$	$1/(1 + 0.15)^3$

or for any year as:

Year	Cash flow		Present worth factor
n	£1 750.00 $\times (1 + d)^n$	$\times$	$1/(1 + i)^n$

This calculation can be simplified by the following adjustment: for $(1 + i)^n$, substitute $(1 + d)^n (1 + e)^n$, where d is the inflation rate as before and e is calculated such that:

$$(1 + i)^n = (1 + d)^n(1 + e)^n$$

giving

$$(1 + e) = \frac{(1 + i)}{(1 + d)}$$

and

$$e\frac{(1 + i)}{(1 + d)} - 1$$

Using this substitution, the calculation of present worth becomes:

Year	Cash flow		Present worth factor
n	£1 750 $\times (1 + d)^n$	$\times$	$\dfrac{1}{(1 + d)^n(1 + e)^n}$

The elements $(1 + d)^n$ cancel and the calculation is reduced to:

$$£1\ 750 \times \frac{1}{(1 + e)^n}$$

The present worth in the original example was calculated by multiplying the cash flow by the present worth factor $1/((1 + i)^n)$, and so the present worth calculated above differs only in the interest rate used. The inflation adjustment has been transferred from the cash flow to the interest rate.

Given the interest rate $i = 0.15$ (15%) and the inflation rate $d = 0.10$ (10%):

$$e = ((1 + 0.15)/(1 + 0.10)) - 1 = 0.045\ 454\ 5 = 4.54\%$$

Therefore, the present worth of Proposal 1, using the interest rate adjusted for inflation, is:

		Proposal 1	
Year	Cash flow (£)	Present worth factors	Present worth (£)
0	8 500	1	8 500.00
1	1 750	0.95652	1 673.90
2	1 750	0.91493	1 601.13
3	1 750	0.87515	1 531.51
4	1 750	0.83710	1 464.93
5	1 750	0.80070	1 401.22
		Total present worth =	£16 172.69

The present worth calculated from the original cash flows and the adjusted interest rate gives a value of £16 172.69, which is the same as that given by Example 4.5, where the interest rate was kept at 15% and the cash flows were adjusted for inflation.

The adjusted interest rate of 4.54% is measuring the interest earned in excess of the inflation rate, and by taking the effect of inflation away from the interest rate it is possible to calculate present worths which allow for the effect of inflation. Calculating the present worth with the adjusted interest rate involves less work than first adjusting the cash flows and then calculating the present worth. Consequently, the method of adjusting the interest rate to allow for inflation is the most commonly employed.

This technique allows for the effects of inflation by reducing the actual interest earned from the apparent rate to an 'effective' or 'real' rate.

In Example 4.6 the interest rate was 15%. This would be the rate that the investor would take to represent the value of money and in all probability would be equated to an interest rate that could be earned in investments deposited elsewhere. The inflation rate used was 10% and the 'effective' interest was calculated at 4.54% for the following reasons. £100 at present-day prices would, at the end of the first year, with inflation at 10%, be equivalent to £110, and the amount required for investment today at 15% to produce £110 in 1 year would be £110 × 0.869 56 = £95.65 (0.869 56 is the present worth factor for 15%). If the £100 required at the end of 1 year were left and the effect of inflation subtracted from the interest rate, then the amount required to be invested today at 4.54% would be £100 × 0.956 52 = £95.65, where 0.956 52 is the present worth factor for 4.54%. Thus, the amount required for investment today would be the same.

Another example would be to assume an interest rate of 15% and an inflation rate of 15%. Since the inflation rate and the rate at which interest is earned are the same, the effective rate becomes zero, as follows:

$$e = ((1 + i)/(1 + d)) - 1 = ((1.15)/(1.15)) - 1 = 0$$

Thus, if £100 at present-day prices were due at the end of 1 year and inflation at 15% made this £115, the amount required to be invested today at 15% to produce £115 in 1 year would be £115 × (0.869 56 = £100 (0.869 56 is the present worth factor for 15%). If the £100 required at the end of 1 year were left and the effect of inflation taken away from the interest rate, then the amount required today at 0% would be £100 × 1.0 = £100, where 1.0 is the present worth factor for 0%. The amount required by both calculations is the same.

This example is particularly noteworthy because, if inflation rates and interest rates are equal, calculating present worths taking account of inflation simply involves summing the cash flows estimated at present prices. The effect of inflation totally eliminates the earned interest.

A final example would be to assume an interest rate of 15%, as before, but an inflation rate of 20% so that the inflation rate is greater than the rate at which interest can be earned. The effective rate then becomes − 4.16% as follows:

$$e = ((1 + i)/(1 + d)) - 1 = ((1.15)/(1.20)) - 1 = - 0.041\ 6$$
$$= - 4.16\%$$

Thus, if £100 at present-day prices were due at the end of 1 year and inflation at 20% would make this £120, the amount required to be invested today at 15% to produce £120 in 1 year would be £120 × 0.869 56 = £104.34, where 0.869 56 is the present worth factor for 15%. If the £100 required at the end of 1 year were left and the effect of inflation taken away from the interest rate, then the amount required today at − 4.16% would be £100 × $1/(1 - 0.041\ 6)^1$ = £100 × 1.043 4 = £104.34, where 1.043 4 is the present worth factor for −4.16%. It is to be noted that the present worth factor had to be calculated from the expression $1/(1 + i)^n$ because negative interest rates are not usually tabulated. Again the amounts required calculated by the two methods are the same.

Thus, this method of adjusting the interest rate is valid for *all* interest and inflation rates and, by producing cash flows estimated on the basis of present-day prices, the effect of inflation at various rates can be easily assessed, using a range of assumed inflation rates.

The following example illustrates the application of this technique to Example 4.2:

EXAMPLE 4.7

Year	Cash flow for the purchase, operating and resale of an item of equipment (£)
0	− 8 500
1	− 1 750
2	− 1 850
3	− 2 000
4	− 2 200
5	− 2 500 + 4 000

The cash flows estimated at present-day prices reflect only the increasing cost of operating the equipment and not increases due to inflation.

The value of money is 15%. The present worth, as shown in Example 4.2, is £13 247.72. If inflation over the next 5 years is estimated at 12%, the effective interest rate would be 2.68%, calculated as follows:

$$e = ((1 + i)/(1 + d)) - 1 = 1.15/1.12 - 1 = 0.026\ 8 = 2.68\%$$

The present worth of the cash flows, allowing for inflation at 12%, is £14 471.42, calculated as follows:

Year	Cash flow (£)	Present worth factors	Present worth (£)
0	− 8 500	1.0	− 8 500.00
1	− 1 750	0.973 9	− 1 704.32
2	− 1 850	0.948 5	− 1 754.73
3	− 2 000	0.923 7	− 1 847.40
4	− 2 200	0.899 6	− 1 979.12
5	− 2 500	0.876 1	− 2 190.25
5	+ 4 000	0.876 1	+ 3 504.40
	Total present worth		− £14 471.42

This £14 471.42 is the present worth of the original cash flows plus the present worth of the additional cash flows that would have to be included for inflation.

4.2.4   Method 4: Varying Inflation Rates

Methods 2 and 3 illustrate how uniform inflation rates can be dealt with, but this leaves the difficulty of coping with varying inflation rates. The most commonly adopted approach in these economic comparisons is to take a long-term view of the interest rate to be used to represent the value of money and to ignore short-term variations. The same argument is usually applied to the assumed inflation rates. It is possible, if required, to cope with varying inflation rates but to do this by adjusting the cash flows as illustrated in Method 3 for uniform inflation rates. The following example based on Example 4.5 demonstrates how cash flows can be adjusted for varying inflation rates. The assumed inflation rates are 10% for years 1 and 2, 12% for year 3 and 14% for years 4 and 5.

EXAMPLE 4.8

		Proposal 1: Inflation adjustments				
Year	Original cash flows (£)	For year 1	For year 2	For year 3	For year 4	For year 5
0	8 500.00					
1	(1 750.00	+ 10%)				
2	((1 750.00	+ 10%)	+ 10%)			
3	(((1 750.00	+ 10%)	+ 10%)	+ 12%)		
4	((((1 750.00	+ 10%)	+ 10%)	+ 12%)	+ 14%)	
5	(((((1 750.00	+ 10%)	+ 10%)	+ 12%)	+ 14%)	+ 14%)

When calculated, these figures become:

	Proposal 1
Year	Cash flows including inflation adjustment (£)
0	8 500.00
1	1 925.00
2	2 117.50
3	2 371.60
4	2 703.62
5	3 082.13

Similarly, the cash flows for Proposal 2 can be adjusted for inflation at follows:

Year	*Proposal 2* Cash flows including inflation adjustment (£)
0	9 500.00
1	1 650.00
2	1 815.00
3	2 032.80
4	2 317.39
5	2 641.83

The present worths of both proposals, calculated at 15%, are £16 412.50 for Proposal 1 and £16 282.14 for Proposal 2. Thus, with this inflation pattern and interest at 15%, Proposal 2 is the more economic.

4.2.5 Note on All Methods

All the examples shown based the comparison on present worth and did not employ EACs. The reason for this is that EAC comparisons involve more calculation than present worth when the annual cash flows are not uniform. When inflation is introduced, the annual cash flows cannot be uniform and it is easier to use present worth calculations.

4.3 VALUATION OF AN ITEM OF EQUIPMENT

The principles of economic comparisons can be employed to place a value on an item of equipment. The term 'value' has several definitions, such as the accountant's value as recorded in the asset register or the market value as determined by how much the item will sell for on the open market. The market value has the most practical meaning, as it is the amount of capital that can be obtained for the equipment item. However, the equipment item may be worth more to the owner than he can obtain by selling the equipment or the market value may be more than the equipment is worth to the owner. To determine either of these values requires the evaluation of worth or value to the owner of the item of equipment. Economic comparisons, based on present worth, can be used to determine the value of the equipment to the owner.

EXAMPLE 4.9

An item of equipment the original purchase price of which was £10 000, and which has been in use for 2 years, has a remaining useful life of 4 years and the estimated running costs at present prices for the next 4 years are £661.25, £766.44, £874.50 and £1 005.68. The estimated resale value at the end of the 4 years is £2 000. Thus, the cash flows are:

Cash flows for
existing plant item

Year	(£)
0	
1	− 661.25
2	− 766.44
3	− 874.50
4	− 1 005.68 + 2 000

Note that there is no cash flow in year zero, because the plant item is already owned.

The present worth of these cash flows, using an interest rate of 10%, is £1 212.45, calculated as follows:

Year	Cash flows for existing equipment item (£)	Present worth factors (10%)	Present worth (£)
0	−	−	−
1	− 661.25	0.909 09	− 601.14
2	− 766.44	0.826 44	− 633.42
3	− 874.50	0.751 31	− 657.02
4	− 1 005.68 + 2 000	0.683 01	+ 679.13
		Total present worth − £1 212.45	

If this item of equipment were not available to the owner, an alternative method of providing the equipment would be necessary and the present worth of the alternative method would have to be calculated and compared with that for the equipment already owned.

Alternative Method 1: Hiring

If the annual hire charge for a similar item of equipment were £2 500 each year for 4 years, then the present worth of hiring this equipment for 4 years, using an interest rate of 10%, would be £7 924.50, calculated as follows:

$$£2\ 500 \times 3.1698 = £7\ 924.50$$

The difference between the two present worths is an estimate of the value of the equipment to the owner. The difference is £7 924.50 − £1 212.45 = £6 712.05, and if the owner were able to sell the equipment for £6 712.05 and to offset this against the cost of hiring, then the present worth of cash flows for hiring would be exactly the same as that for already owning the equipment. If the owner were able to sell the equipment for more than £6 712.05, it would be more economic to do so and to hire the equipment. If £6 712.05 could not be realised by selling the existing equipment, it would be more economic to retain the existing item of equipment.

If the effects of inflation were to be introduced, this could be achieved by adjusting the interest rates as explained previously, in the section dealing with inflation adjustments.

Alternative Method 2: Buying New Equipment

If a similar item of equipment were available for purchase, then the cash flows for purchase, operating and resale would have to be estimated as in the next example.

EXAMPLE 4.10

The cash flows estimated at present prices for purchasing, operating and reselling after 4 years for a similar item of equipment are:

Year	Cash flow for purchase, operating and resale for new item of equipment (£)
0	− 11 000.00
1	− 400.00
2	− 460.00
3	− 529.00
4	− 608.35 + 4 000

The present worth of these cash flows, using an interest rate of 10%, is calculated as follows:

Year	Cash flows for new item of equipment (£)	Present worth factors (10%)	Present worth (£)
0	− 11 000.00	1.0	− 11 000.00
1	− 400.00	0.909 09	− 363.64
2	− 460.00	0.826 44	− 380.16
3	− 529.00	0.751 31	− 397.44
4	− 608.35 + 4 000	0.683 01	+ 2 316.53
		Total present worth	− £ 9 824.71

The difference between the present worth of keeping the existing equipment and replacing immediately with a new item of equipment is:

$$£9\ 824.71 - £1\ 212.45 = £8\ 612.26$$

Thus, if the owner could sell the existing equipment for £8 612.26 and offset this against the cost of the new equipment, the present worth of acquiring the new equipment would be the same as keeping the existing item of equipment. If the owner could sell the item of equipment for more than £8 612.26, then it would be more economic to sell and replace: if £8 612.26 could not be realised from the sale of the equipment, it would be more economic to retain the equipment.

If the effects of inflation were to be introduced, this could be achieved by adjusting the interest rates as explained previously, in the section dealing with inflation adjustments.

4.4   VALUATIONS FOR LONGER LIVES

The above valuations have been calculated from the useful life of the existing item of equipment. This was estimated at 4 years, the equipment already being 2 years old and having a total life of 6 years. If the need for the equipment extended for, say, 20 years, the comparison would also need to be extended for that time period. Thus, the cost of keeping the existing equipment and its subsequent replacements must be compared with the cost of immediate replacement and subsequent replacements. The replacements will all be estimated at the same costs as the immediate replacements, as all cash flows are estimated at present prices and the effects of inflation incorporated separately.

EXAMPLE 4.11

The cash flows for keeping the existing item of equipment and subsequent replacements for 20 years are as follows:

Cash flows

Year	Running costs of existing equipment (£)	Resale of existing equipment (£)	Purchase of replacements (£)	Running costs of replacements (£)	Resale of replacements (£)
0	–				
1	– 661.25				
2	– 766.44				
3	– 874.50				
4	– 1 005.68	+ 2 000.00	– 11 000.00		
5				– 400.00	
6				– 460.00	
7				– 529.00	
8				– 608.35	
9				– 699.60	
10				– 804.54	+ 2 000
11			– 11 000.00	– 400.00	
12				– 460.00	
13				– 529.00	
14				– 608.35	
15				– 699.60	
16			– 11 000.00	– 804.54	+ 2 000
17				– 400.00	
18				– 460.00	
19				– 529.00	
20				– 608.35	+ 4 000

The cash flows for immediate replacement with a new item of equipment and replacements for 20 years are:

		Cash flows	
Year	Purchase price (£)	Running costs (£)	Resale (£)
0	− 11 000		
1		− 400.00	
2		− 460.00	
3		− 529.00	
4		− 608.35	
5		− 699.60	
6	− 11 000	− 804.54	+ 2 000
7		− 400.00	
8		− 460.00	
9		− 529.00	
10		− 608.35	
11		− 699.60	
12	− 11 000	− 804.54	+ 2 000
13		− 400.00	
14		− 460.00	
15		− 529.00	
16		− 608.35	
17		− 699.60	
18	− 11 000	− 804.54	+ 2 000
19		− 400.00	
20		− 460.00	+ 8 000

To compare these cash flows, the present worth of both must be calculated, and the present worth of the immediate replacement can be calculated as follows, using an interest rate of 10%:

Year	Cash flow (£)	Present worth factor (10%)	Present worth (£)
0	− 11 000.00	1.0	− 11 000.00
1	− 400.00	0.909 09	− 363.64
2	− 460.00	0.826 44	− 380.16
3	− 529.00	0.751 31	− 397.44
4	− 608.35	0.683 01	− 415.51
5	− 699.60	0.620 92	− 434.40
6	− 804.54 + £2 000.00	0.564 47	+ 674.80
		Total present worth	− £12 316.35

Thus, the cash flows for the immediate and subsequent replacements can be represented as shown below:

Year	Cash flow (£)
0	− 12 316.35
6	− 12 316.35
12	− 12 316.35
18	− 11 000.00
19	− 400.00
20	− 460.00 + 8 000

The £12 316.35 at years 0, 6 and 12 represents all the cash flows for the replacements purchased in those years, and the present worth for the immediate and subsequent replacements up to 20 years is £24 115.92, calculated as follows:

Year	Cash flow (£)	Present worth factor (10%)	Present worth (£)
0	− 12 316.35	1.0	− 12 316.35
6	− 12 316.35	0.564 47	− 6 952.21
12	− 12 316.35	0.318 63	− 3 924.36
18	− 11 000.00	0.179 85	− 1 978.35
19	− 400.00	0.163 50	− 65.40
20	− 460.00 + £8 000	0.148 64	+ 1 120.75
		Total present worth	− £24 115.92

The present worth of keeping the existing equipment until the end of its useful life and its subsequent replacement can be calculated first by representing the cash flows as shown below:

Year	Cash flows (£)
0	− 1 212.45
4	− 12 316.35
10	− 12 316.35
16	− 11 000.00
17	− 400.00
18	− 460.00
19	− 529.00
20	− 608.35 + 4 000

The £1 212.45, taken from Example 4.9, represents all the cash flows for the existing item of equipment.

The £12 316.35 in years 4 and 10 is used to represent all the cash flows for the replacement in those years. Consequently, the present worth for keeping the existing plant item until the end of its useful life

and subsequent replacements up to 20 years is £16 511.14, calculated
as follows:

Year	Cash flows (£)	Present worth factor (10%)	Present worth (£)
0	− 1 212.45	1.0	− 1 212.45
4	− 12 316.35	0.683 01	− 8 412.19
10	− 12 316.35	0.385 54	− 4 748.45
16	− 11 000.00	0.217 62	− 2 393.82
17	− 400.00	0.197 84	− 79.14
18	− 460.00	0.179 85	− 82.73
19	− 529.00	0.163 50	− 86.49
20	− 608.35 + £4 000	0.148 64	+ 504.13
		Total present worth	− £16 511.14

The difference between these two present worths is £24 115.92 −
£16 511.14 = £7 604.78. If the owner could sell the existing equipment for £7 604.78 and this amount could be offset against the cost of
the immediate replacement, the present worth of the immediate and
subsequent replacements would be the same as keeping the existing
item. If the owner could sell the existing equipment for more than
£7 604.78, the acquiring of an immediate replacement would be more
economic. If £7 604.78 could not be realised by the sale of the existing
equipment, then keeping the existing equipment would be more
economic.

   If the effects of inflation were to be introduced, this could be
achieved by adjusting the interest rates as explained previously, in
the section dealing with inflation adjustments.

The above calculation was based on an overall duration of 20 years, and
the present worths were comparable because they were calculated for
equal periods of time. The life of the equipment is 6 years. With immediate
and subsequent replacements the 20 years is made up by disposing of the
replacement in year 18 after 2 years, and by adjusting the resale value of
this last replacement. Similarly, the last replacement in the case of keeping
the existing equipment and replacing from year 4 onwards occurs in year 16,
and again the resale value of this last replacement has been adjusted.

EXAMPLE 4.12

An alternative to assuming a finite cut-off of 20 years, as in the last
example, is to assume that the replacements will continue in perpetu-

ity. The cash flow for immediate and subsequent replacements to infinity can be represented as follows:

	Cash flows representing
Year	replacements to infinity (£)
0	− 12 316.35
6	− 12 316.35
12	− 12 316.35
18	− 12 316.35
24	− 12 316.35
.	.
.	.
.	.
∞	∞

The £12 316.35 is the present worth of purchasing, operating and reselling the new equipment. This present worth was calculated previously as part of Example 4.11. If this equipment is replaced every 6 years, this amount recurs every 6 years to infinity.

The present worth of a sum recurring every year, starting at the end of the first year, is given by the factor $((1 + i)^n − 1)/(i(1 + i)^n)$ which is the uniform series present worth factor given in the Appendix.

The present worth of a sum recurring not every year but every y years, starting in y years, is given by the factor $1/((1 + i)^y − 1)$. Thus, if the sum recurring every y years were x, the present worth would be

$$x \times \left(\frac{1}{(1 + i)^y − 1} \right)$$

and if the sum x occurred in year 0, the total present worth of the whole series would be

$$x + x \left(\frac{1}{(1 + i)^y − 1} \right)$$

which reduces to

$$\frac{x}{1 − (1/(1 + i)^y)}$$

Thus, substituting £12 316.35 for x, 6 years for y and 10% for i, the present worth of the series starting in year 0 and recurring to infinity is

$$\frac{£12\ 316.35}{1 - 0.564\ 47} = £28\ 278.99$$

The 0.564 47 can be taken from tables, since the element $1/(1 + i)^y$ is the expression for the present worth factor.

The amount calculated, £28 278.99, represents the amount that would be required today which, if invested at 10%, would produce £12 316.35 every 6 years forever. This can be checked as follows. First take away the initial £12 316.35; this leaves £15 962.64 for investment. In 6 years this £15 962.64 would increase to £15 962.64 × 1.771 56 = £28 278.99. This £28 278.99 is made up of the original capital £15 962.64, and interest earned in those 6 years of £12 316.35. The factor 1.771 56 is the compound amount factor for 10% and is taken from tables in the Appendix. Thus, the sum of £12 316.35 can be used every 6 years to provide replacement equipment. This £12 316.35 is the interest earned on the capital of £15 962.64 and is entirely used up every 6 years. The capital of £15 962.64 remains undepleted and can go on producing £12 316.35 every 6 years indefinitely.

The £12 316.35 is the present worth of purchasing, operating and reselling the equipment item, and £28 278.99 is the present worth of purchasing, operating and reselling every 6 years in perpetuity. This present worth is sometimes called the capitalised cost.

To arrive at a valuation of the existing equipment, it is necessary to calculate the capitalised cost of the cash flows relating to the scheme whereby the existing equipment is retained — that is, the present worth of the existing equipment and its replacements to infinity.

Taking the cash flows from Example 4.11 for keeping the existing equipment and its replacements, these can be represented as follows:

| | Cash flows for keeping existing equipment and its replacements | | |
| | | | |

Year	Running costs of existing equipment (£)	Resale (£)	Replacements (£)
0	–		
1	– 661.25		
2	– 766.44		
3	– 874.50		
4	– 1 005.68	+ 2 000	– 12 316.35
10			– 12 316.35
16			– 12 316.35
22			– 12 316.35
28			– 12 316.35
.			.
.			.
.			.
∞			∞

The cash flows in years 0–4 representing the running costs of the existing equipment and the resale value have a present worth of £1 212.45, as calculated in Example 4.9.

The £12 316.35 every 6 years has a present worth of £28 278.99, as calculated in Example 4.11. However, in this set of cash flows the first of the £12 316.35 occurs in year 4 and the cash flows for keeping the existing equipment and its replacements to infinity can be represented as follows:

Year	Cash flows for keeping the existing equipment and its replacements (£)
0	– 1 212.45
4	– 28 278.99

The present worth of these cash flows is £20 527.28, calculated thus:

$$£1\ 212.45 + (£28\ 278.99 \times 0.683\ 01) = £20\ 527.28$$

Given an interest rate of 10%, £20 527.28 is the amount required for investment today to give £1 212.45 in year 0, enough for the existing equipment, and £28 278.99 in year 4, enough for the replacements at £12 316.35 every 6 years until infinity. Thus, £20 527.28 is the capitalised cost of keeping the existing equipment and then replacing it in perpetuity.

The difference between the capitalised cost of keeping the existing equipment and its immediate replacement is the value to the owner of the existing equipment. The difference is £28 278.99 − £20 527.28 = £7 751.71. This £7 751.71 is the amount that could be compared with the market value when considering disposal. The value calculated here is only slightly different from that calculated for replacements up to 20 years, because the present worth factors become smaller with increasing time and the effect of cash flows beyond 20 years on such calculations is small.

If the effects of inflation were to be incorporated in this calculation, this would be achieved by adjusting the interest rates as explained previously, in the section dealing with inflation adjustments.

4.5   DETERMINING REPLACEMENT AGE

The factors that determine the economic replacement age of equipment are the purchase price, the operating costs and the resale value. The purchase price is relevant, because the equipment must be kept long enough to warrant the investment. The operating costs usually increase with ageing equipment and it is, therefore, important not to keep the equipment too long. Also, as these operating costs increase, the resale value declines. The purpose of an economic analysis is to find the balance between these. As the comparisons are being made between keeping equipment for 1, 2, 3 or 4 years, etc., the use of present worth presents the difficulty of equalising the lives of the comparisons. This may be achieved by considering replacements to infinity. As the operating costs are not usually uniform, the use of equivalent annual costs is not easily applied either. Thus, neither present worth nor equivalent annual costs offer any real advantage in this comparison. The method presented here is based on equivalent annual costs.

4.5.1   Determining Replacement Age Using Equivalent Annual Costs

The concept of equivalent annual costs is explained in Example 4.3.

EXAMPLE 4.13

The purchase price of a small concrete batching unit is £25 000. The operating costs, based on the estimated annual average hours of use, are £1 000 in the first year when manufacturers' warranties operate,

Table 4.1   Determining replacement age by use of EACs

	A	B	C	D	E	F	G
Year	Purchase price (£)	Capital re-covery factors for 15%	EAC of purchase price (A × B) (£)	Running costs (£)	Present worth factors at 15%	Present worth of running costs (D × E) (£)	Sum of present worth of running costs (ΣF) (£
0	25 000				1.0		
1		1.150	28 750	1 000	0.869	869.00	869.00
2		0.615	15 375	1 500	0.756	1 134.00	2 003.00
3		0.437	10 925	1 875	0.657	1 231.88	3 234.88
4		0.350	8 750	2 250	0.571	1 284.75	4 519.63
5		0.298	7 450	2 625	0.497	1 304.63	5 824.26
6		0.264	6 600	3 000	0.432	1 296.00	7 120.26

Note: The minimum in column N, £6 955.35, is the minimum EAC.

and £1 500 in the second year, rising by £375.00 per year thereafter. The resale values are as follows:

Year	Predicted resale values (£)
1	22 500
2	20 000
3	18 750
4	15 000
5	10 000
6	6 250

The calculations to determine the equivalent annual cost (EAC) of keeping this equipment for 1, 2, 3 and 4 years, etc., are set out in Table 4.1.

Most of the entries in Table 4.1, which calculate the EACs for 1, 2, 3, etc., years, are self-explanatory.

Column C shows the EAC of the purchase price. For example, if the equipment were kept for 4 years, the EAC of the purchase price would be £8 750 per year.

Column F is the present worth of the running costs. In year 4 the running costs for that year were £2 250, and the present worth of this amount is £1 284.75: i.e. the amount required to be invested in year 0 to provide £2 250 in year 4 is £1 284.75.

Column G is the running total of the present worths of the running costs so £3 234.88 is the sum of the present worths for years 1, 2 and 3. Thus, if £3 234.88 were invested in year 0 at 15%, it would provide enough to pay for the running costs of years 1, 2 and 3.

H	I	K	L	M	N
EAC of present worth of running costs (B × G) (£)	EACs of purchase price and running costs (C + H) (£)	Resale value (£)	Present worth of resale (K × E) (£)	EACs of resale (B × L) (£)	EACs of purchases running and resales (I − M) (£)
999.35	29 749.35	22 500	19 552.50	22 500.00	7 249.35
1 231.84	16 606.84	20 000	15 120.00	9 298.80	7 308.04
1 413.64	12 338.64	18 750	12 318.75	5 383.29	6 955.35
1 581.87	10 331.87	15 000	8 565.00	2 997.75	7 334.12
1 735.62	9 185.62	10 000	4 970.00	1 481.06	7 704.56
1 879.74	8 479.74	6 250	2 700.00	712.80	7 766.94

Column H is the EAC of the present worth of the running costs. Taking the £3 234.88 for year 3 from column G and converting it to an annual sum, using the capital recovery factor, gives £1 413.64, and the running costs of £1 000, £1 500 and £1 875 for years 1, 2 and 3 are now converted to a uniform series of £1 413.64 per year. The calculations in columns F, G and H are devices to convert the varying operating costs to a uniform series.

Column I is the EAC for the purchase price added to the equivalent annual cost for the running costs.

Column L is the present worth of the resale value. £12 318.75 in year 3 is the present worth of the resale value of £18 750 in year 3.

Column M is the EAC of the resale value. Thus, £5 383.29 for year 3 is the equivalent annual cost of £12 318.75 from column L.

Column N is the EAC of the resale subtracted from the equivalent annual costs of the purchase price and running cost. This gives the net EAC for purchase, operating and resale for years 1–6.

Thus, the minimum in column N is the minimum EAC and the best replacement age by this economic criterion. The minimum is £6 955.35 in year 3.

The effects of inflation could be incorporated by adjusting the interest rates that were used to calculate the present worth, but not the interest rate used to calculate the EACs.

BIBLIOGRAPHY

See list at end of Chapter 5.

Chapter 5
Equipment Profitability

5.1 MEASURING PROFITABILITY

The economic analyses described in Chapter 4 have related to schemes whereby only the expenditure was considered. The purchase price and running or operating costs were not offset in the calculations against any revenue, and the only monies returning to the investor were from resale. Thus, the previous analyses were confined to determining whether one course of action — say the purchase of one particular item of equipment — was more economic than another course of action, the purchase of an alternative item. What will be considered now is the situation where the equipment generates a revenue. The simplest situation to imagine is where the equipment is purchased and hired out so that the owner has capital expenditure, operating costs, revenue and resale value. The analysis required now is not simply whether one item is more economic than another but whether the item is earning an adequate return on the invested capital. That is, whether the return on capital derived from owning and hiring out equipment is better than could be obtained from less risky investment elsewhere or, if the capital to buy the equipment is borrowed, whether the return is greater than the cost of capital as measured by the interest on the capital. The capital made available by the company to purchase the equipment should earn at least the minimum return expected by the company. This analysis, therefore, requires that the rate of return be measured and the most widely used method is known as the 'internal rate of return', 'yield' or 'discounted cash flow [DCF] yield'. All these are names for the same measure of profitability.

The interest factors used in this chapter are taken from the Appendix.

5.1.1   Discounted Cash Flow Yield

Calculating the DCF yield requires first that the net cash flows be calculated. The net cash flows are the sum of the cash flows relating to the investment project. It is useful in calculating the net cash flows to construct a cash flow tree or model as shown in Figure 5.1, a cash flow tree for a simple purchase, operating, hire and resale model. This shows the cash flows broken down into three levels. The first level is the net cash flow. It is from this net cash flow that the DCF yield will be calculated. The next level represents the three general categories of 'capital cost', 'operating cost'

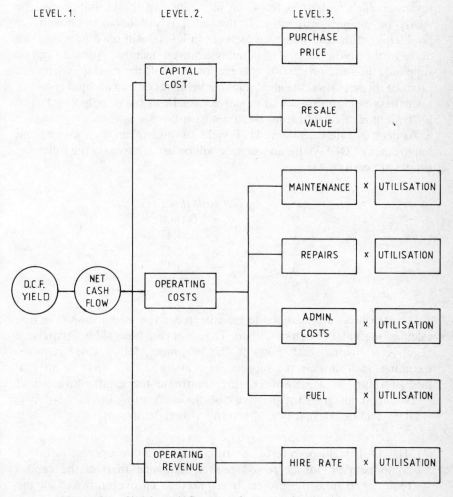

Figure 5.1   Simple cash flow tree for an equipment hire operation

and 'operating revenue'. These three general categories apply to most cash flow trees. It is the next level, level 3, that determines the unique features of the operation or investment being modelled, and breaks down the capital costs to purchase price and resale. The operating costs are broken down into estimates for 'maintenance', 'repairs', 'administration' and 'fuel'. The estimate for the purchase price is a single sum occurring in year 0 and the estimate for the resale value is a single sum occurring in the year when the resale takes place. The estimates for the operating costs are not single sums but estimates of cost for each year of operation. In more detailed models these estimates of cost may be made for each month. All the operating costs are shown tied to the utilisation, as clearly a relationship exists between the utilisation and these costs. Similarly, the operating revenue which is derived from the hire rate and the utilisation will be yearly or monthly estimates. If this level is not detailed enough for producing estimates, then each element in level 3 will have to be broken down further. When a level of detail is achieved such that estimates can be supplied, then the process of aggregation through the cash flow tree will produce the net flows, the algebraic sign being taken into account (positive cash flows being income and negative cash flows being outgoings). The DCF yield is calculated from these net cash flows.

To demonstrate calculating DCF yield and to explain the meaning and importance of DCF yield, an example will be used, based on the following simple net cash flows:

Year	Net cash flows (£)
0	− 1 000.00
1	+ 315.47
2	+ 315.47
3	+ 315.47
4	+ 315.47

These cash flows are assumed to have been derived from a cash flow tree calculation similar to that described. These net cash flows show a pattern of negative, outgoing, cash flows in the beginning, followed by positive, incoming, cash flows in the subsequent years. The DCF yield or internal rate of return is a measure of the return on the capital invested of − £1 000.00, given by the positive cash flows occurring in years 1–4.

DCF yield or internal rate of return has two definitions:

(1) The DCF yield or rate of return is the *maximum* interest rate that could be paid for borrowed capital, assuming that all the capital required to fund the project is acquired as an overdraft and all the positive cash flows are used to repay this overdraft.

(2) The DCF yield or rate of return is the interest rate which, if used to discount a project's cash flow, will give a net present value (or worth) of zero.

The second of these definitions is more usually employed in calculating the yield, and is also the definition that has given rise to the name 'DCF yield'. It is also the more difficult to understand at first reading, so the example will be calculated in the first instance using the first definition.

Calculation of DCF yield by the first definition is a process of trial and error. An interest rate is assumed and tested to determine whether it is the maximum: if it is above the maximum, a new trial with a smaller interest rate is used; if it is below the maximum a new trial with a larger interest rate is used until the maximum is found, eventually by interpolation, if necessary.

Using an interest rate of 12% as the first trial rate, the test as to whether this is the maximum is shown in the following table:

Year	Net cash flow (£)	Interest paid on borrowed capital at 12% (£)	Borrowing account (£)
0	− 1 000.00	−	− 1 000.00
1	+ 315.47	− 120.00	− 804.53
2	+ 315.47	− 96.54	− 585.60
3	+ 315.47	− 70.27	− 340.41
4	+ 315.47	− 40.85	− 65.79

In year 0 the amount borrowed was − £1 000.00 and the interest on this during year 1 was − £120.00. Thus, at the end of year 1 the borrowing account was the original − £1 000.00 together with the interest of − £120.00 offset by the income of + £315.47, leaving − £804.53 in the borrowing account. Repeating this calculation until the end of the project indicates that there is − £65.79 left in the account. Thus, this project could not have paid interest at 12% on the borrowed capital, because to do so would require £65.79 from other sources. Thus, 12% is greater than the maximum interest rate that this project could support and the trial and error process continues with a smaller interest rate. Suppose that the second guess is 8%: the test as to whether this is the maximum is shown in the table below:

Year	Net cash flow (£)	Interest paid on borrowed capital at 8% (£)	Borrowing account (£)
0	− 1 000.00	−	− 1 000.00
1	+ 315.47	− 80.00	− 764.53
2	+ 315.47	− 61.16	− 510.22
3	+ 315.47	− 40.82	− 235.57
4	+ 315.47	− 18.85	− 61.05

Proceeding through the calculations as before, the amount left in the account at the end of the project is + £61.05. Thus, this project could have paid more for its borrowed capital than 8%, which is therefore less than the maximum interest rate.

Given that 12% is greater than the maximum interest rate and 8% is less than the maximum interest rate, the maximum clearly lies between the two. Therefore, the next reasonable guess would be 10%. The trial to determine whether 10% is the maximum is:

Year	Net cash flow (£)	Interest paid on borrowed capital at 10% (£)	Borrowing account (£)
0	− 1 000.00	−	− 1 000.00
1	+ 315.47	− 100.00	− 784.53
2	+ 315.47	− 78.45	− 547.51
3	+ 315.47	− 54.75	− 286.79
4	+ 315.47	− 28.68	− 0.00

Performing the calculation as before shows that an amount of zero would be left in the borrowing account. Thus, 10% is the maximum interest rate that could be paid for the borrowed capital in this project, and by definition (1) 10% is the DCF yield.

Measuring this *maximum* that *could* be paid for the borrowed capital is a way of measuring how much the project cash flows are producing. Measuring the amount that can be taken away (by interest charges) shows how much the project is producing. Also, this measure (the DCF yield or rate of return) is directly comparable with the cost of capital, the cost of capital being the weighted average of the costs of all sources of capital used by the company. If the company were paying more for its capital than 10%, the project would not be satisfactory, because it would not be yielding more than borrowing the capital is costing.

The method more commonly employed in calculating the DCF yield is according to definition (2). Calculating the DCF yield by the second definition is also a trial and error process which also requires an assumed interest rate and a trial to determine whether the assumed interest rate gives an NPV (net present value or worth) of zero. If the calculated NPV is negative the assumed interest rate is too large and a smaller one is assumed and the NPV is recalculated. If the calculated NPV is positive the assumed interest is too small. The process is repeated until the interest rate which gives a zero NPV is found — by interpolation, if necessary.

Using 9% as a first trial, the NPV is calculated and found to be positive; a larger interest rate of 11% is then used to produce a negative NPV. Interpolation produces the interest rate which gives an NPV of zero and is found to be 10%, as before.

Year	Net cash flows (£)	Present worth factors for 9% (1st trial)	Present worth (£)	Present worth factors for 11% (2nd trial)	Present worth (£)
0	− 1 000.00	1.0	− 1 000.00	1.0	− 1 000.00
1	+ 315.47	0.917 43	+ 289.42	0.900 90	+ 284.21
2	+ 315.47	0.841 68	+ 265.52	0.811 62	+ 256.04
3	+ 315.47	0.772 18	+ 243.60	0.731 19	+ 230.67
4	+ 315.47	0.708 42	+ 223.49	0.658 73	+ 207.81
			+ £22.03		− £21.27

By interpolation: the interest rate which gives an NPV of zero

$$= 9\% + (11\% - 9\%) \times \left(\frac{22.03}{22.03 - (- 21.27)} \right)$$
$$= 9\% + 2\% \times 0.508$$
$$= 9\% + 1.01\% = 10.0\%$$

The present worth factors are taken from tables or calculated as $1/(1 + i)^n$. This 10.0% has exactly the same meaning as the 10.0% calculated by the first definition. The trial and error process that produces the interest rate that gives an NPV of zero can be understood by examining the graph of NPV against interest rates as shown in Figure 5.2. Any minor discrepancy between the interpolated value and 10% is due to interpolation which assumes that the graph in Figure 5.2 is linear between 9% and 11%. It is, in fact, slightly curved.

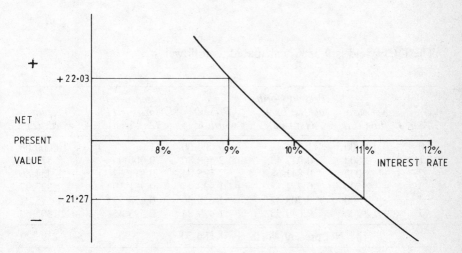

Figure 5.2   Graph of net present value against interest rate

To understand further the meaning of this calculated interest rate which has been given the name 'DCF yield' or 'internal rate of return' consider this question: if £1 000 were invested at 10%, what regular income could be taken each year for the next 4 years?

The capital recovery factor for 10% and 4 years is 0.315 47. This factor is taken from the tables in the Appendix. Thus, the income that could be taken each year for 4 years from £1 000 invested at 10% is:

$$£1\ 000 \times 0.315\ 47 = £315.47$$

This £315.47 is the same as the net cash flow in the example used.

Given a capital sum and an interest rate, the income can be calculated. Also, given the capital sum and the income, the interest rate that would produce that income from that capital sum can be calculated. This interest rate is called the DCF yield or the internal rate of return.

This example was based on uniform revenues, but the searching techniques for DCF yield will work equally well for non-uniform revenues, as shown in the next example, in which the net cash flows are:

Year	Net cash flows (£)
0	− 10 000
1	+ 2 800
2	+ 3 000
3	+ 2 500
4	+ 2 300
5	+ 2 200

The DCF yield is 9.35%, calculated as follows:

Year	Net cash flow (£)	Present worth factors for 9% (1st trial)	Present worth (£)	Present worth factors for 11% (2nd trial)	Present worth (£)
0	− 10 000	1.0	− 10 000.00	1.0	− 10 000.00
1	+ 2 800	0.917 43	+ 2 568.80	0.900 90	+ 2 522.52
2	+ 3 000	0.841 68	+ 2 525.04	0.811 62	+ 2 434.86
3	+ 2 500	0.772 18	+ 1 930.45	0.731 19	+ 1 827.97
4	+ 2 300	0.708 42	+ 1 629.36	0.658 73	+ 1 515.07
5	+ 2 200	0.649 93	+ 1 429.84	0.593 45	+ 1 305.59
		Net present worth +	£83.50	−	£393.99

Interpolation:

$$\text{DCF yield} = 9\% + (11\% - 9\%) \times \left(\frac{83.50}{83.50 - (-393.99)} \right)$$
$$= 9\% + (2\% \times 0.175)$$
$$= 9\% + 0.35\%$$
$$= 9.35\%$$

The yield or rate of return is the most widely used measure of profitability.

5.1.2 Other Measures of Profitability

Yield is not the only measure of profitability: others in use are net present value, payback period and average annual rate of return.

Net present value is used to determine whether a proposed project yields at least the minimum return specified by the company. The NPV is calculated for the net cash flows, using the minimum rate of return required, this rate representing the cost of capital to the company. If the NPV is positive, it follows that the yield is above the minimum and the project is worthy of further consideration. If the NPV is negative, the yield is less than the minimum and the project can be rejected without further analysis.

The payback period is the time taken to repay the original capital invested. The payback period is not very useful without predetermining what a satisfactory payback period should be. If a company usually expects payback periods of 1 year or 8 months it would clearly be unhappy with proposals that incurred payback periods of 3 or 4 years. In general, the shorter the payback period the more profitable the project, provided that the project continues to have positive net revenues for a number of years after the payback period.

The average annual rate of return is all the returns (positive cash flows) for the project averaged over the number of years the project lasts and expressed as a percentage of the invested capital. Like the payback period, the average annual rate of return is not very useful without first determining what a satisfactory rate should be. An average annual rate of return of 33% and a payback period of 3 years are similar. The higher this rate of return the more profitable the project.

The payback period or the average annual rate of return are never used on their own as measures of profitability but always in conjunction with other measures, such as NPV or even DCF yield.

5.2   VARYING HIRE RATES AND YIELD

If a cash flow tree such as that illustrated in Figure 5.1 is constructed to represent an equipment hire operation, then the effects of the key elements in that cash flow tree can be studied by substituting a range of values for the key variables and determining the different net cash flows and, hence, the different yields. Two major variables in an equipment hire operation are the hire rate and the utilisation factor. So, if a range of five hire rates is used with a range of seven utilisation factors from 50% to 110%, different net cash flows can be calculated for each of the combinations, making a total of 35. For each of the net cash flows, a yield can be calculated. Results of this type are best presented graphically, as shown in Figure 5.3. This graph is known as a sensitivity chart and the analysis performed is known as a sensitivity analysis, as it is displaying the sensitivity of yield to both the hire rate and the utilisation factor.

This family of curves illustrates two obvious points: the greater the hire rate, and the greater the utilisation factor, the greater the rate of return. But while these points are obvious, this type of graph quantifies the increase in rate of return for any assumed increase in hire rate or utilisation factor. For example, for hire rate number 3, £10 per hour, the graph illustrates that the plant has to be on hire for 77% of the normal maximum

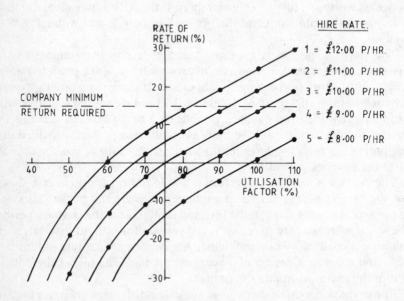

Figure 5.3   Sensitivity of rate of return to the hire rate and the utilisation of an item of equipment

usage before it would show a positive return, and to record a return equal to the company's minimum at this hire rate, the equipment would need to be on hire all the available normal working time plus some overtime working. Adding the company's minimum return required to the graph reveals the combination of usage rate and hire rate that would produce this return or more. The hire rate eventually adopted would be chosen for marketing reasons. These sensitivity graphs help to assess whether the market hire rate is likely to produce an adequate return on capital.

5.3   YIELD AND THE EFFECT OF CORPORATION TAX AND CAPITAL ALLOWANCES

If a company trades profitably, then it will be subject to corporation tax. Currently, since the budget of 1984 corporation tax rate in the UK is 35%, except for the smaller companies. There is a delay between the company engaging in trade, declaring a profit and paying tax. This lag in paying tax varies and depends on the relationship between the company's accounting year and the fiscal year. The lag could vary from 9 months to 18 months. In the example which follows and which is used to demonstrate the effect of corporation tax on yield, two simplifying assumptions have been made: the first is that we are dealing with a large company and, hence, the corporation tax rate is 35%; the second is that a tax lag of 1 year operates. These assumptions, although simplifying, are valid.

A system of capital allowances also exists and must be included in this example. At the time of writing and since the 1984 budget, UK tax legislation allows capital allowances for industrial plant, including construction plant, to be taken at 25% written down. That is, 25% of the investment can be set against tax initially and in the following year 25% of the remaining capital investment can be set against tax. This process repeats in subsequent years; consequently, the amount of the allowance reduces.

Thus, in our large company, corporation tax is being paid at 35%, a tax time-lag of 1 year exists, and capital allowances for investment in plant are 25% written down.

If the company proposes to invest in a large item of earthmoving equipment, the net cash flows estimated at present-day prices are as given in Table 5.1.

Corporation tax and capital allowances calculations are shown in Table

Table 5.1 Estimated cash flows

Year	Purchase price and resale value (£)	Operating costs (£)	Operating revenue (£)	Net revenue (£)
0	50 000			
1		20 000	45 000	25 000
2		22 000	45 000	23 000
3		24 000	40 000	16 000
4		26 000	40 000	14 000
5		28 000	36 000	8 000
6	(15 000)	30 000	35 000	5 000

Total net revenue = £91 000
Capital invested = £50 000
Resale value   = £15 000

5.2, where:

Column A shows the original purchase price of £50 000 and the resale value of £15 000.

Column B shows the net revenues.

Column C shows the corporation tax that would be due on these net revenues if no capital allowances were in operation. The corporation tax in column C is time-shifted by 1 year to reflect the tax time-lag. Thus, the £8 750 of calculated corporation tax shown in year 2 is calculated on the net revenues shown in year 1 in column B.

Column D shows the capital allowance of £50 000, calculated at 25% written down of the purchase price, and a balancing allowance of £15 000, calculated on the resale value. The balancing allowance is to compensate for the £15 000 of capital returning to the company from the resale. Having already received a capital allowance of £43 325.79 (the aggregate of column D) on the original £50 000, the company now finds that the capital invested was only £35 000, made up of £50 000 − £15 000: hence the balancing allowance.

Column E is the tax saved by the capital allowance, calculated as the capital allowance times the tax rate. The tax due on the balancing allowance is also calculated at the balancing allowance times the tax rate.

Column F is the tax paid, whichi s column C, less the savings in column E or plus the tax due in column E.

Column G is the net cash flows after tax. The capital invested in year 0 is offset by the tax saving that year. Year 1 is the £25 000 net revenue from column B plus the tax saving from column F. The explanation for this is that the tax saving can only be taken if the profits and, hence, the tax due are sufficient to accommodate that saving. If this project were the only project the company had, then the profits and the tax in year 1 would be insufficient and the tax saving would have to be rolled forward to be absorbed in later years. However, it was assumed that this was a large company, and it is further assumed that this large company has sufficient

Table 5.2 Calculation of net of tax cash flows

	A	B	C	D	E	F	G
Year	Purchase price and resale (£)	Net revenue (£)	Corporation tax due on previous year's net revenues 35% (£)	Capital allowance 25% written down (balancing allowance) (£)	Tax saved (or owed) on allowances (£)	Tax paid (£)	Net cash flow after tax (£)
0	50 000			12 500.00	4 375.00	(4 375)	− 45 625.00
1		25 000		9 375.00	3 281.25	(3 281.25)	− 28 281.25
2		23 000	8 750	7 031.25	2 460.94	6 289.06	16 710.94
3		16 000	8 050	5 273.44	1 845.70	6 204.30	9 795.70
4		14 000	5 600	3 955.07	1 384.27	4 215.72	9 784.28
5		8 000	4 900	2 966.30	1 038.20	3 861.80	4 138.20
6	15 000	5 000	2 800	2 224.73	778.65	2 021.34	17 978.66
7			1 750	(8 325.79)	(2 914.03)	4 664.03	− 4 664.03

profits to accommodate this saving in years 0 or 1 of this project. Because this project has given rise to those savings (cash that would otherwise have left the company), the savings are credited to this project. Years 2–5 are the net revenues from column *B*, less the tax in column *F*. Year 6 is the resale value plus the net revenue from column *B*, less the tax in column *F*. Year 7 is the tax from column *F*.

The introduction of corporation tax and capital allowances greatly reduces the profit but it also distorts the cash flows. The cash flow in year 1, for example, is larger than before tax considerations, while the others are much smaller. The calculations in Tables 5.3 and 5.4 compare the yields obtained before tax and after tax.

Table 5.3   Calculation of yield before tax

Year	Cash flow (£)
0	− 50 000
1	+ 25 000
2	+ 23 000
3	+ 16 000
4	+ 14 000
5	+ 8 000
6	+ 20 000

Table 5.4   Calculation of yield after tax

Year	Cash flow (£)
0	− 45 625.00
1	+ 28 281.25
2	+ 16 710.94
3	+ 9 795.70
4	+ 9 784.28
5	+ 4 138.20
6	+ 13 739.06

The calculations were performed using the computer program listed in *Modern Construction Management*, by Harris and McCaffer (Blackwell Scientific). The results were:

Payback period	2.13 years
DCF yield	31.43%

Before calculating the yield after tax, it is necessary to adjust the cash flows to remove the negative cash flow in year 7. The method of calculation explained previously is known as the single rate calculation and is not suitable for cash flows that have negatives anywhere other than the starting years. If used on cash flows with large negative cash flows at the end, it could produce more than one interest rate, giving a zero net present value. To overcome this, provision is made for the negative cash flow by taking a suitable amount from the previous year's positive cash flow. The amount set aside is the negative cash amount discounted by 1 year. The rate used to discount the negative cash flow is called the 'earning rate' and represents the return that can be obtained from a safe investment. Thus, using an earning rate of 10%, the negative cash flow of £4 664.03 in year 7 can be discounted to £4 239.60 (£4 664.03 × 0.909 0, where 0.909 0 is the present worth factor taken from tables). If this £4 239.60 is deducted from the positive cash flow of £17 978.66 in year 6, then 6's cash flow becomes £13 739.06. Thus, the £4 239.6 removed from year 6 provides for the negative cash flow in year 7 and the cash flows are now suitable for a single rate DCF yield calculation as before. For a fuller explanation see *Modern Construction Management*.

The computer program listed in the Investment Appraisal Exercise in *Modern Construction Management* was used for the calculations. The results were:

Payback period	2.06 years
DCF yield	27.1%

Thus, although the corporation tax removed considerable amounts from this investment project, the yield rate of return was reduced only from 31.43% to 27.1%. The reason for the yield remaining so high is the distortion in the cash flows brought about by the 100% capital allowance in the first year. This gave the net cash flows a relatively large positive cash flow in year 1. This occurred only because the tax saving arising from the capital allowance was taken and credited to this project. If the company were unable, owing to lack of profit, to take immediately the benefit of the tax savings given by the capital allowances and had to delay taking the savings until enough profits were available, the yield after tax would fall. Thus, the benefits of capital allowances are at their greatest when the company is trading profitably enough to use the capital allowances immediately.

The example shown here was presented in the form of appraising a proposed investment, the cash flows of which were estimated at present prices. Taxation, of course, would be calculated on the actual cash flows as they occurred. These actual cash flows would be subjected to inflation and would reflect the effects of inflation. The effects of inflation on yield calculations are explained in the next section.

5.4   YIELD CALCULATIONS AND INFLATION

If cash flow estimates are made at present-day prices, the yield calculated on these cash flows will not reflect the effects of inflation. An example would be the purchase and hiring of an item of equipment, as shown below (Table 5.5).

Table 5.5   Cash flow estimates for purchasing and hiring an item of equipment

		Cash flows		
Year	Investment (£)	Operating costs (£)	Operating revenue (£)	Net cash flows (£)
0	10 000			− 10 000
1		3 000	6 000	+ 3 000
2		3 500	7 000	+ 3 500
3		4 000	7 000	+ 3 000
4		4 500	7 000	+ 2 500

The yield of this project is 8.0%, calculated as shown in Table 5.6.

Table 5.6   Calculation of yield

Year	Net cash flow (£)	Present worth factors at 9%	Present value (£)	Present value factors at 7%	Present value (£)
0	− 10 000	1.0	− 10 000.00	1.0	− 10 000.00
1	+ 3 000	0.917 43	+ 2 752.29	0.934 57	+ 2 803.71
2	+ 3 500	0.841 68	+ 2 945.88	0.873 43	+ 3 057.00
3	+ 3 000	0.772 18	+ 2 316.54	0.816 29	+ 2 448.87
4	+ 2 500	0.708 42	+ 1 771.05	0.762 89	+ 1 907.23
		Net present value	− £214.24	Net present value	+ £216.81

By interpolation:

$$\text{Yield} = 7\% + (9\% - 7\%) \times \left(\frac{216.81}{216.81 - (-214.24)} \right)$$
$$= 7\% + 1\%$$
$$= 8.00\%$$

As the estimates were based on present-day prices, this 8.00% yield does not include any effects of inflation.

Including inflation at the uniform rate of 10% per year on both the operating costs and the operating revenues, the cash flows adjusted for inflation are as shown in Table 5.7.

Table 5.7 Cash flows adjusted for inflation

Year	Investment (£)	Cash flows						Revised net cash flows (£)
		Operating costs (£)	Inflation adjustment (£)	Revised operating cost (£)	Operating revenue (£)	Inflation adjustment (£)	Revised operating revenue (£)	
0	10 000.00							− 10 000.00
1		3 000.00	300.00	3 300.00	6 000.00	600.00	6 600.00	+ 3 300.00
2		3 500.00	735.00	4 235.00	7 000.00	1 470.00	8 470.00	+ 4 235.00
3		4 000.00	1 324.00	5 324.00	7 000.00	2 317.00	9 317.00	+ 3 993.00
4		4 500.00	2 088.45	6 588.45	7 000.00	3 248.70	10 248.70	+ 3 660.25

Table 5.8 Calculation of yield on cash flows adjusted for inflation

Year	Revised net cash flows (£)	Present value factors at 19%	Present value (£)	Present value factors at 18%	Present value (£)
0	− 10 000.00	1.0	− 10 000.00	1.0	− 10 000.00
1	+ 3 300.00	0.840 33	+ 2 773.09	0.847 45	+ 2 796.59
2	+ 4 235.00	0.706 16	+ 2 990.59	0.718 18	+ 3 041.49
3	+ 3 993.00	0.593 41	+ 2 369.49	0.608 63	+ 2 430.26
4	+ 3 660.25	0.498 66	+ 1 825.22	0.515 78	+ 1 887.88
		Net present value	− £ 41.61	Net present value	+ £ 156.22

The revised net cash flows are in fact simply the original net cash flows inflated at 10% per year. The yield of these cash flows, which now include inflation, is calculated as shown in Table 5.8.

By interpolation:

$$\text{Yield} = 18\% + (19\% - 18\%) \times \left(\frac{156.22}{156.22 - (-41.61)} \right)$$
$$= 18\% + 1\% \times 0.79$$
$$= 18.8\%$$

This yield, calculated on the revised cash flows, is larger than the yield calculated on the original cash flows, because the revised positive cash flows themselves were larger. These cash flows were larger because of the inflation included in them. The £3 993 revised net cash flow in year 3 does not buy any more goods than the £3 000 original net cash flow. The difference between the two amounts is cancelled by inflation. The £993 more in the revised cash flows just compensates for the inflation at 10% per year. Thus, although the revised positive net cash flows are larger and, hence, the yield is larger at 18.8%, the investor is no better off, because the extra monies acquired are absorbed by inflation. Thus, 8% on the original *uninflated* cash flows is equivalent to 18.8% on the revised *inflated* cash flows. To distinguish between these two rates of return, the rate of return calculated on the estimates which did not include inflation, the original *uninflated* cash flows, is called the *real rate of return*. The rate of return calculated on the estimates which included inflation, the revised *inflated* cash flows, is called the *apparent rate of return*.

The relationship between these two rates of return is through the inflation rate. The apparent rate of return is the real rate of return, increased by the inflation rate as follows:

$$(1 + a) = (1 + r)(1 + d)$$

where a is the apparent rate of return, r is the real rate of return and d is the inflation rate.

Substituting the values calculated from Tables 5.6 and 5.8 for a and r ($a = 0.188$ (18.8%) and $r = 0.08$ (8%)) into the expression gives a value for d, the inflation rate:

$$(1 + 0.188) = (1 + 0.08)(1 + d)$$
$$\therefore (1 + d) = 1.188/1.08$$
$$\therefore d = 1.10 - 1 = 0.10 = 10\%$$

Given the value of the calculated real rate of return at 8% and the calculated apparent rate of return at 18.8%, the expression estimates that the inflation rate is 10%. Since the inflation rate included in the estimates was 10%, this serves as a check on the above explanation.

As most proposed investments are appraised on estimates based on present-day prices, the rate of return calculated and used to judge a proposal's viability is normally the real rate of return. Since the cash flows recorded as projects proceed are normally based on the transactions that occur at current prices, the cash flows determined usually have inflation included as a matter of course. Thus, the yield calculated on these recorded cash flows will be the apparent rate of return. It is, therefore, important to distinguish whether the cash flows are based on constant, year 0, prices or current prices when interpreting the yield calculated.

The situation used to explain the relationship between real rate, apparent rate and the inflation rate was one where the equipment owner was allowed to increase his revenue at the same rate as inflation. However, if this were not the case, the apparent rate would be less than 18.8% but the inflation would still be 10% and the *achieved* real rate of return would be less than the *estimated* real rate of return of 8%.

The following example illustrates this. If the project originally described in Table 5.5 had been executed and completed, the cash flows could have been recorded. Throughout the duration of the project, inflation had been 10% per year. Public spending cuts, high interest rates and a general recession had made it impossible for equipment hirers to raise the hire rates in line with inflation, although rates had increased to some extent. Thus, the operating revenue, although increasing, had not kept pace with inflation. The recorded cash flows are as shown in Table 5.9. The apparent rate of return calculated on these recorded cash flows in 12.85%, as calculated in Table 5.10.

Table 5.9   Cash flows recorded during the execution of the project

Year	Investment (£)	Operating costs (£)	Operating revenue (£)	Net cash flow (£)
0	10 000.00			− 10 000.00
1		3 300.00	6 435.00	+ 3 135.00
2		4 235.00	8 057.09	+ 3 822.09
3		5 324.00	8 747.50	+ 3 423.50
4		6 588.45	9 569.75	+ 2 981.30

Table 5.10 Calculated apparent rate of return on recorded cash flows

Year	Recorded net cash flows (£)	Present value factors at 13%	Present value (£)	Present value factors at 12%	Present value (£)
0	− 10 000.00	1.0	− 10 000.00	1.0	− 10 000.00
1	+ 3 135.00	0.884 95	+ 2 774.32	0.892 85	+ 2 799.08
2	+ 3 822.09	0.783 14	+ 2 993.23	0.797 19	+ 3 046.93
3	+ 3 423.50	0.693 05	+ 2 372.66	0.711 78	+ 2 436.78
4	+ 2 981.30	0.613 31	+ 1 828.46	0.635 51	+ 1 894.65
			− £31.33		+ £177.44

By interpolation:

$$\text{Apparent rate of return} = 12\% + (13\% - 12\%) \times \left(\frac{177.44}{177.44 - (-31.33)} \right)$$

$$= 12\% + 1\% \times 0.85$$

$$= 12.85\%$$

Thus, the achieved apparent rate of return is 12.85%, but as inflation during this time was 10% per year, the real rate of return achieved is 2.59%, calculated as follows:

$$(1 + a) = (1 + r)(1 + d)$$
$$(1 + 0.128\ 5) = (1 + r)(1 + 0.1)$$
$$\therefore (1 + r) = 1.128\ 5/1.1$$
$$\therefore r = 1.025\ 9 - 1 = 0.025\ 9 = 2.59\%$$

Thus, because the income or revenue was restrained from increasing at the rate of inflation but costs were rising at the rate of inflation, the real rate of return was reduced to 2.59%.

Other examples could be apparent rates of return of 10%, inflation rates of 10% and real rates of return of zero, or an apparent rate of 8%, an inflation rate of 10% and a real rate of return of − 1.8%.

Inflation evidently reduces the real rate of return unless prices are allowed to rise to compensate for its effects. This reduction in the real rate of return could drag the achieved real rate of return below the cost of capital. In other words, the capital could well be costing more than the project is yielding. In such a case the project is uneconomic, and if such projects are sustained, the company will become bankrupt. The first noticeable effect probably will be that the company cannot replace its equipment and will either go on using ageing equipment or reduce its fleet.

The construction equipment hire industry is particularly vulnerable to the effects of inflation. This is because the construction industry is largely an industry that does not create its own demand and is dependent on public works and public spending for a substantial portion of its workload. The rest of the industry's workload depends on the private sector being willing to invest. If the cost of money, i.e. interest rates, is high, the private sector is discouraged from investing and the demand for construction work declines. Since in times of rising inflation governments cut public spending while also increasing interest rates, the demand for construction declines accordingly. As the demand declines, prices for construction work falls and the equipment hire industry cannot raise prices in line with inflation. As a result, the gap between operating costs and revenue is squeezed, with a consequential fall in the return on capital invested in the equipment.

BIBLIOGRAPHY

Coombs, W. E. and Palmer, W. J. (1984). *The Handbook of Construction Accounting and Financial Management*, 3rd edn, McGraw-Hill

Greig Middleton & Co. Ltd. *Investment in Plant Hire 1988/89*, published by Greig Middleton for private circulation.

Harris, F. C. and McCaffer, R. (1986). *Worked Examples in Construction Management*, BSP Professional Books

Harris, F. C. and McCaffer, R. (1989). *Modern Construction Management*, 3rd edn, Blackwell Scientific

Lumby, S. (1984). *Investment Appraisal*, 2nd end, Van Nostrand Reinhold

Pilcher, R. (1983). *Project Cost Control*, Blackwell Scientific

Pritchard, W. E. (1986). *Corporation Tax*, 9th edn, Pitman

Samuels, J. M. and Wilkes, F. M. (1986). *Management of Company Finance*, 4th edn, Van Nostrand Reinhold

Chapter 6
Equipment Acquisition

6.1 METHODS OF ACQUISITION

Much of Chapters 4 and 5 dealt with the economic analyses that help decide whether or not to acquire an item of equipment, based mainly on the question: Does this proposed acquisition offer the opportunity to earn an adequate rate of return and which of the possible items of equipment is the most economic? So far the question of how the item of equipment should be acquired has not been considered. There is a tendency in very large companies for the two decisions as to whether to acquire and how to acquire to be taken separately, the specialist to acquire and the finance responsible for the decision as to whether to acquire and the finance directors being responsible for the decision as to how to acquire. In smaller companies these two decisions often get merged. Major methods of acquisition are reviewed in this chapter, and the relative advantages of each method are highlighted.

The decision to acquire an asset should be made for both technical and economic reasons. The profitability of the proposal should be evaluated by calculating the expected rate of return and comparing it with the cost of capital. The decision as to how to acquire the asset can then be regarded as a financial one.

The major methods of acquisition can be classified as purchase, leasing or hiring. The major factors that influence the decision as to which is the more advantageous are:

(1) Tax legislation, which allows 25% written down capital allowances against the purchase of construction plant.
(2) The profit flows of the acquiring company, which determine whether these allowances can be turned into tax savings benefiting the company

immediately or rolled forward until later years, thus becoming devalued.

(3) The acquiring company's cash flows, which determine what money is available for plant acquisition.

(4) The acquiring company's gearing ratio (borrowed capital/equity capital), which influences the amount of further borrowing possible.

6.1.1   Purchase

Outright purchase is simply payment of the purchase price by the acquiring company to the supplier. This involves the acquiring company in a large cash payment very early, before the equipment acquired has earned any revenue. However, outright purchase provides the acquiring company with capital allowances of 25% written down of the purchase price of the equipment. If the acquiring company's profit flows are sufficient, these allowances can produce a saving of 35% of the allowances. Thus, an item of plant worth £10 000 would produce capital allowances of £2 500 immediately and tax savings of £875.00 (£2 500 × 35%, 35% being the corporation tax rate). This tax saving is most valuable but it is only available if the profit flows in the company are £2 500 or more.

If the cash is available from within the company's own resources or even from an overdraft, this form of acquisition is probably the cheapest, provided that the capital allowances can be used to produce the tax saving immediately. If the capital allowances cannot be used immediately because the company's profit flows are inadequate, then the capital allowances can be rolled forward until sufficient profit flows are available. In this situation the benefit of the capital allowances and derived tax savings become devalued, in simple present worth (or value) terms, and in these circumstances outright purchase may not be the cheapest method of acquisition.

Outright purchase places the title of the equipment immediately with the acquiring company. This means that it becomes an asset over which the company has full control; which it can use in negotiating finance; which it can use anywhere, including overseas; and which it can dispose of to produce cash from its resale value.

Other methods of purchase include credit sale and hire purchase.

A *credit sale* is a sale in which the acquiring company takes the ownership or title of the plant item immediately but the purchase price is paid in instalments. These instalments include the purchase price plus any financing charges the vendor makes. Credit sales, like outright purchase, attract capital allowances immediately and can be used in the same way as outright purchase.

Hire purchase and *leasing* are significantly different in the treatment of tax and therefore must be considered quite separately. Hire purchase is a contract whereby the acquiring company pays a regular hire charge and, at some predetermined point after payment of a proportion of the agreed hire charges, the acquiring company buys the equipment for a nominal sum. This facility to purchase distinguishes the hire purchase contract from leasing, which, under UK tax legislation, does not permit the acquiring company to purchase the leased equipment. Hire purchase also attracts the capital allowances as though the equipment were purchased outright. Thus, in terms of tax savings, hire purchase has the same advantages as outright purchase.

Both hire purchase and credit sales are likely to require deposits, but these deposits are much less than the whole purchase price and therefore in cash flow considerations these forms of acquisitions are less demanding than outright purchase. However, the interest charges included in hire purchase contracts are likely to be greater than those the acquiring company would pay on an overdraft. Thus, if the hire purchase method of acquisition is compared with outright purchase, outright purchase is cheaper in most cases, the capital allowances available in both cases being the same.

6.1.2 Leasing

The leasing method of acquisition is different in concept from the previous methods outlined. The difference is that the ownership or title of the equipment remains the property of the leasing company (the lessor), and the acquiring company (the lessee) never becomes the owner. The acquiring company (the lessee) only acquires the *use* of the equipment in return for payments or rentals but never becomes the owner. While this method is more common in the leasing of property, it is also used in the acquisition of the use of capital equipment such as construction equipment. Although there is a plethora of leasing arrangements, they all adhere to the basic principle that the lessor is the owner and the lessee is the user of the equipment. There are two broad categories of lease — the finance lease and the operating lease.

The *finance lease* is normally arranged through leasing companies who have no particular interest in the equipment and offer no technical support, but merely arrange the lease. The lessee pays the lessor payments or rentals for the use of the equipment acquired. The equipment is usually supplied by a third party — the equipment manufacturer or manufacturers' agent from whom the equipment will be bought by the leasing company. The payments or rentals for this type of lease will be divided into two parts — the primary period and the secondary period. The duration of the primary period is dependent on the useful of the equipment, but is 2–5 years for most construction equipment. Payments during this primary period are calculated by the leasing company to include the capital cost of the equipment, less any allowance for the resale value, plus the leasing company's additions for their own overheads, interest charges and profit. The capital cost will also reflect the capital allowances that the leasing company will be able to claim against the leasing company's tax. Thus, the capital allowances due do not go to the acquiring company using the equipment, as in purchasing, because the lessee is not the owner. The capital allowances are claimed by the owner who purchased the equipment, and that is the leasing company. The advantages the lessee gains are from the amount of this benefit that is passed on by the leasing company through the payments or rentals. Payments for any secondary period are negotiated but could be relatively small, as the leasing company has already recovered all costs and profit during the primary period. The using company is not permitted by tax legislation in the UK to buy the equipment or to sell it, although the leasing company, which is unlikely to have any interest or ability to use the equipment, is free to sell it and may share the proceeds with the user company who had leased the equipment. The conditions of such a lease agreement tend to prevent the company that leases the equipment from cancelling the agreement during the primary period. The contract is also likely to specify responsibility for insurance, maintenance, servicing and repairs. All these conditions are designed to protect the owner's property for the duration of the primary period. The leasing company's contribution is simply that of providing finance.

The outgoings of the lease to the user company, the lessee, are the lease payments or rentals: the lessee does not have to find the purchase price, or deposits, as in hire purchase or credit sales, but may have to pay about 3 months' rental in advance. Thus, the cash flow for leasing is less difficult to arrange. Some leasing companies have been known to arrange uneven lease payments which are negotiated to match the cyclical use of equipment. Thus, the leasing of earthmoving equipment which may be idle during December, January and February could well benefit from such uneven repayment schemes.

As described the lessee does not receive the capital allowances directly: these go to the owner — the leasing company — and the benefit is passed on via the lease payments. This benefit is available to the lessee regardless of his profit flows. Thus, whether the lessee (the construction company) has large profits able to benefit from capital allowances or small profits unable to benefit from capital allowances is no longer relevant, as it is in purchasing. Thus, leasing in terms of costs when compared with the purchasing option is more advantageous to a company when its profit flows are small and it is unable to use the capital allowances to generate tax savings immediately; or when the company is rapidly expanding and its investment programme has created such a total of capital allowances that, even though it is profitable, it is unable to use all these capital allowances. In addition to the capital allowances, the lease payments themselves are normal trading expenses and therefore deductible from revenue before calculating tax due. Thus, there is some tax saving on the lease payments as well as any benefit from the capital allowances the leasing company may have passed on when calculating the lease payment.

Another feature of finance leases is that the security of the lease may well be only the asset itself and the rest of the company's borrowings against the company's owned assets may not be affected by a leasing arrangement. Therefore, a company that is already 'highly geared', i.e. has high borrowings in relation to equity or shareholders' capital, may find leasing attractive. Although the leased asset may not show on the balance sheet, the company has nevertheless committed itself to payments and in practical terms is just as vulnerable as if it had increased its borrowings.

Thus, a finance lease is likely to be more advantageous to a construction company when its profit flows do not generate the full benefit of capital allowances for purchasing and/or when the company cash flow situation is unable to provide funds for purchase, or when the company is unable to undertake further borrowings to purchase equipment.

The *operating lease* is normally arranged with manufacturers or suppliers who offer such a service as part of the marketing of their products. Again such leases are likely to have a non-cancellable primary period, but the duration and costs may be quite different from those of finance leases because the leasing company, being the manufacturer, has a different interest in the equipment. For example, the supplying company may have use for the equipment itself or a well-developed secondhand leasing market. In these circumstances an operating lease may be cheaper than a finance lease. The capital allowances would, as with finance leases, go to the leasing company who own the equipment.

6.1.3   Hiring

Hiring and leasing are sometimes regarded as similar and for equipment items that are on hire for long periods the difference between the two methods of acquisition may be unclear. An example of such a long-term hire is the contract hire arrangement for vehicles. The payments are similar to those in leases, and the owner of the vehicle — the hire company — claims the capital allowances. However, contract hire arrangements can involve the hire company supplying the vehicles in providing repair and maintenance, whereas finance leases do not involve the leasing company in providing these services. Thus, there are distinctions. Short-term hire of construction equipment is not usually regarded as leasing, and the company hiring the equipment pays an hourly, weekly or monthly rate for the equipment. The period of hire may well be as short as 1 week or 1 month and therefore the construction company is not committed to a long primary period, as it would be in a finance lease. The use of such short-term hire is, of course, widespread in the construction industry, and within the industry there exists a very well-developed equipment hire industry to serve this market. Many construction companies who own their own equipment run the equipment division as a subsidiary offering external hire to other companies and internal hire to their own construction division, which may be the parent company or another company within the same holding group. Thus, the construction companies, themselves the users of the equipment, are well used to hiring either internally or externally. This conveniently separates the problems of equipment acquisition, determining adequate hire rates and marketing of equipment to ensure adequate utilisation from the work of construction. All these costs are simply reflected in the hire rate to the construction company.

However, it is worth noting that if a subsidiary equipment hire company sets its internal hire rates at strictly economic levels, the construction division may be able to obtain ostensibly better external hire rates from outside companies. These external rates may not have been chosen for economic reasons, or the external hire company may have lower fixed costs and overheads and therefore be able to offer cheaper hire rates. If the external hire is chosen in preference to an apparently more expensive internal hire rate, the parent company may be damaged by the loss of hire income and by the unrecovered part of the fixed costs, which still must be met even while the equipment is idle.

Thus, most of the acquisition of construction equipment, whether by purchase or lease, is by the equipment hire companies, which are either companies specifically set up to provide the hire service to construction companies or equipment subsidiaries of construction companies.

6.2   COMPARISON BETWEEN LEASING AND PURCHASING

The key to deciding whether leasing is more advantageous than purchasing in cost terms is the use companies can make of the capital allowances. For example, ignoring tax considerations, if the purchase price of an item of equipment were £12 000 and the lease payments were £1 450 per quarter for 3 years, the cost of these two cash flows could be compared on present worth terms, using an interest rate that represented the value of money to the company. Table 6.1 illustrates this, using an interest rate of 15%.

Table 6.1 Comparison between leasing and purchasing, ignoring tax considerations

Year	Quarter	Period	Purchasing (£)	Leasing (£)
0		0	− 12 000	
	1	1		− 1 450
1	2	2		− 1 450
	3	3		-- 1 450
	4	4		− 1 450
	1	5		− 1 450
2	2	6		− 1 450
	3	7		− 1 450
	4	8		− 1 450
	1	9		− 1 450
3	2	10		− 1 450
	3	11		− 1 450
	4	12		− 1 450

$$\text{Present worth of purchasing} = - £12\,000$$
$$\text{Present worth of leasing} \;\; = - £1\,450 \times 9.634\,96$$
$$= - £13\,970.69$$

Note: The factor 9.634 96 is the present worth factor for a uniform series calculated as

$$\frac{(1 + i)^n - 1}{i(1 + i)^n}$$

where n is the number of periods, 12, and i is the interest rate per quarter. The interest rate per annum was given as 15%. Thus, the interest rate per quarter is calculated, using the compound relationship, as

$$(1 + i_{\text{quarter}})^4 = (1 + i_{\text{year}})^1$$
$$\therefore (1 + i_{\text{quarter}})^4 = (1.15)^1$$
$$\therefore i_{\text{quarter}} = \sqrt[4]{(1.15)} - 1 = 0.035\,5 = 3.55\%$$

Thus, at an interest rate of 15% the purchase alternative is cheaper. In fact, the interest rate in this example would need to be 27.2% per year or 6.2% per quarter before the leasing costs would just equal the purchase price. The situation is influenced, however, when tax considerations are included.

The leasing versus purchasing comparison is a financial appraisal and usually the discount rate used in these comparisons represents the cost of borrowing. The cost of borrowing is the nominal interest rate, less tax for the profitable company.

Table 6.2 shows the cash flows for outright purchase, using the 25% written down capital allowances immediately and assuming a tax lag of 1 year to represent the delay in meeting a tax liability or, as in this case, recording a tax saving. The tax saving is treated the same as an inward cash flow, since this has the same effect as a saving that prevents an outward cash flow. The tax saving is calculated on a tax rate of 35%. The net present value at 15% is − £9 527.37, as shown in Table 6.2.

Table 6.2 Cash flows and NPV calculation for outright purchase including tax saving from capital allowances used immediately

Year	Purchase price (£)	Capital allowances (£)	Tax saving (£)	Net cash flow (£)	Present worth factors (15%)	Present worth (£)
0	− 12 000	3 000.00	1 050.00	− 10 950.00	1.0	− 10 950.00
1		2 250.00	787.50	+ 787.50	0.869 56	684.78
2		1 687.50	590.62	+ 590.62	0.756 14	446.59
3		1 265.62	442.97	+ 442.97	0.657 51	291.26
				Net present value		− £9 527.37

Note: Capital allowances and tax savings after year 3 have been ignored. To include these would reduce the net present value calculated and make the purchase option appear even more economic.

Table 6.3 shows the net cash flows for the leasing alternative. No tax savings from capital allowances are shown, as these would already be reflected in the lease payment charged to the acquiring company. However, tax savings derived from the lease payments are shown, as these are normal trading expenses and are deducted from revenue before tax. A tax lag of 1 year is assumed before the tax reductions are included. The tax rate is 35%. The net present value calculated at 15% is − £9 940.33.

Thus, as with the case of Table 6.1, which ignores tax considerations, when tax considerations are included and the capital allowances available in outright purchase are used immediately, the outright purchase option is cheaper than the lease option, as the net present values of − £9 527.37 for outright purchase and − £ 9 940.33 for leasing illustrates.

However, the situation changes if there is a delay in taking the tax savings from the capital allowances created by the purchase.

Table 6.4 calculates the net present value for a delay to year 2 before taking the tax savings. This delay could be caused by profit flows being insufficient to use the capital allowances immediately.

With a delay to year 2 before the tax savings become effective, the difference between purchasing and leasing narrows, and the net present value of leasing remains at − £9 940.33, whereas the net present value of

Table 6.3   Net cash flow for the leasing alternative including tax considerations

Year	Quarter	Period	Lease payment (£)	Tax reduction (£)	Present worth factors (15%)	Present worth of tax reduction (£)
1	1	1	− 1 450			
	2	2	− 1 450			
	3	3	− 1 450			
	4	4	− 1 450			
2	1	5	− 1 450			
	2	6	− 1 450			
	3	7	− 1 450			
	4	8	− 1 450	+ 2 030	0.756 14	+ 1 534.96
3	1	9	− 1 450			
	2	10	− 1 450			
	3	11	− 1 450			
	4	12	− 1 450	+ 2 030	0.657 51	+ 1 334.75
4	4	16		+ 2 030	0.571 75	+ 1 160.65

The tax reduction is calculated as 35% of a year's lease
payments and delayed 1 year (4 × £1 450.00 × 35%).

Present value of tax reductions	+ £ 4 030.36
Present value of lease payments, from Table 6.1	− £13 970.69
Net present value:	− £ 9 940.33

Table 6.4   NPV of outright purchase with a delay to year 2 in using capital allowances

Year	Purchase (£)	Tax saving (£)	Net cash flow (£)	Present worth factors (15%)	Present worth (£)
0	− 12 000		− 12 000	1.0	− 12 000
1					
2		+ 2 428.12	+ 2 428.12	0.756 14	1 836.00
3		+ 442.97	+ 442.97	0.657 51	291.26
			Net present value:		− £ 9 872.74

Note: The tax savings are calculated by rolling forward the unused allowances
calculated in Table 6.2.

purchasing becomes − £9 872.74. Any further delay in taking the tax
savings will make leasing a cheaper alternative, as Table 6.5 illustrates by
adding a further year delay before taking the tax saving.

Thus, with a delay to year 3 before the tax savings become effective, the
leasing option becomes more economic as the net present value of

Table 6.5 NPV of outright purchase with a delay to year 3 in using capital allowances

Year	Purchase (£)	Tax saving (£)	Net cash flow (£)	Present worth factors (15%)	Present worth (£)
0	− 12 000		− 12 000	1.0	− 12 000
1					
2					
3		2 871.09	+ 2 871.09	0.657 51	1 887.77
			Net present value:		− £10 112.23

Note: The tax savings are calculated by rolling forward the unused allowances as calculated in Table 6.2.

purchasing becomes − £10 112.23, whereas the net present value of leasing remains at − £9 940.33.

A final illustration shows the capital allowances spread over years 2 and 3 to represent the ability to take some tax savings in year 2 and the remainder in year 3. This is shown in Table 6.6. Again this illustrates that delays in taking the tax savings can lead to the leasing option becoming more economic, as the net present value for outright purchase is − £9 970.65, while the net present value for leasing is − £9 940.33.

Table 6.6 NPV of outright purchase with capital allowances used in years 2 and 3

Year	Purchase (£)	Tax saving (£)	Net cash flow (£)	Present worth factors (15%)	Present worth (£)
0	− 12 000		− 12 000	1.0	− 12 000
1					
2		1 435.54	+ 1 435.54	0.756 14	1 085.46
3		1 435.55	+ 1 435.55	0.657 51	943.89
			Net present value:		− £ 9 970.65

In all the lease versus purchase examples shown above, a tax lag of 1 year was adopted to represent the delay in meeting a tax liability. This 1 year lag could vary and the lag itself influences this lease versus purchase comparison. However, the point is made in these examples that any delay in using the capital allowances that are created by purchasing shifts the balance of the economic comparison in favour of leasing. Thus, the cost comparison between leasing and buying is determined by the company's profit flows.

The other factors to be considered are:

(1) The company's cash flow, since leasing makes fewer demands than purchasing.
(2) The company's ability to raise the capital to purchase, since this depends on the extent of the company's existing borrowings.
(3) The availability of alternative uses for the investment funds available.
(4) The loss of flexibility, due to commitments, to make lease payments annually.
(5) The restrictions placed on the using company in using the equipment.

Thus, the resolution of the lease versus purchase issue is not simply an economic comparison but involves these other factors as well.

BIBLIOGRAPHY

Coombs, W. E. and Palmer, W. J. (1984). *The Handbook of Construction Accounting and Financial Management*, 3rd edn, McGraw-Hill

Harris, F. C. and McCaffer, R. (1989). *Modern Construction Management*, 3rd edn, Blackwell Scientific

Pizzey, A. (1985). *Accounting and Finance, A Firm Foundation*, Holt, Rinehart and Winston

Pritchard, W. E. (1986). *Corporation Tax*, 9th edn, Pitman

Samuels, J. M. and Wilkes, F. M. (1986). *Management of Company Finance*, 4th edn, Van Nostrand Reinhold

Chapter 7
Systematic Selection of Equipment

7.1 TECHNICAL EVALUATION

The purchase of an item of construction equipment requires a high capital outlay, and the consequences of misjudging the potential earnings of the machine over a number of years could have a dire effect on future profits. Thus, unless it can be shown that the equipment will yield a rate of return at least as good as making an alternative investment, it should not be purchased at all. In this event either leasing or hiring may be a more economical option.

However, even when the financial and other economic factors have been adequately satisfied regarding the purchase, because of the many alternative choices of manufacturer now available for many items of equipment, the final decision is likely to be influenced by the merits of small but important differences offered with each make. It is this aspect of the purchase — namely the technical evaluation — which is dealt with in this chapter.

The decision process follows a systematic approach originally developed by US consultants Kepner and Tregoe. The method forces the manager into a sequence of actions and helps to highlight the relevant factors. In this way the many separate judgements needed for an examination of many facts can be weighted and ranked accordingly, and the best buy for the least cost can be chosen.

7.2 THE MAIN FEATURES OF DECISION MAKING

Dixon describes decision making as follows:

Decision making is compromise. The decision maker must weigh value judgements that involve economic factors, technical practicabilities, scientific necessities, human and social considerations, etc. To make a 'correct' decision is to choose the one alternative from among those that are available which best balances or optimises the total value, considering all the various factors.

Using this definition, Kepner and Tregoe established seven essential factors in decision making (Figure 7.1).

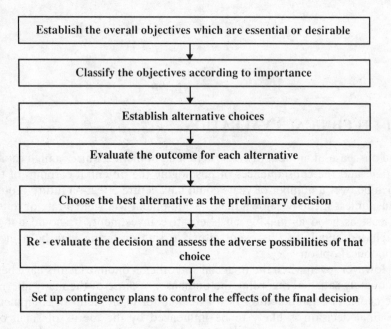

Figure 7.1   Seven essential factors in decision making

EXAMPLE 7.1: CRANE EVALUATION

A company has recently obtained a construction contract which will involve the use of a 40-tonne-capacity, strut-boom, crawler-mounted crane. A 9 year working life can be assumed, which indicates that an attractive rate of return may be achieved by purchasing the crane. The crane must be capable of conversion to a dragline/grabbing crane and cost less than £41 000.

The plant manager is required to undertake a technical evaluation of the alternative makes of crane available in the market.

The Method

Step 1: Set Objectives Which Are Essential and Desirable

The problem as defined is far too vague to establish the precise requirements, and some discussion with the site manager, planning engineer and even the operator is required. From these further enquiries the precise features of the machine may be ascertained: these will be called the 'objectives', and examples are listed A–G, 1–18 in Figure 7.2.

Step 2: Classify Objectives According to Importance

A careful examination of the listed objectives will reveal that some are mandatory, some are important but not essential, while others would be desirable and useful if available.

From this list of objectives, the mandatory objectives are set aside and called *musts*. The *musts* set the limits which cannot be violated, so makes of machine which do not meet the fundamental requirements are quickly recognised and eliminated early in the analysis.

The remaining objectives are labelled *wants*, and some may be more important than others. The next step, therefore, is to rank the importance of the *want* objectives. This is achieved by attaching a ranking expressed on a suitable scale. In practice, a 1–5 or 1–10 system is suitable, with the importance of the objective increasing with numerical value. This part of the procedure usually requires much subjective assessment and may vary according to an individual's preference, but at least a record stands for future reference and reconciliation.

The *want* objectives are classified into two groupings:

(1) The performance results expected from the machine – here called the *technical element*.
(2) The demands made upon resources of labour, money, materials, time – here called the *economic element*, as associated with costs and servicing.

Step 3: Develop the Alternative Choices

In this arbitrary example, three (entirely fictitious) manufacturers are located, making a crane which may be suitable for the specified duties. These are: (1) Harris 46T; (2) HFC 416; (3) Sirac Highlift.

At this stage only the *musts* are considered. Consequently, the field of choice may be narrowed to a small number, saving much time and money.

Objectives			Alternatives	
Ref	Musts		Harris 46T	Go/No
A	Crawler mounted machine		Yes	✓
B	At least 150′ boom as lift crane		160′ max main	✓
C	Capable of conversion to dragline/excavator		Yes	✓
D	Delivery less than 10 weeks		6/8 weeks	✓
E	Max purchase price £41 000		Yes	✓
F	Conform to B.S. and statutory regulations		Yes	✓
G	40 tonnes minimum lifting capacity		Yes	✓

No.	Wants	Ele-ment	Ranking	Information	Rating	Weighted score = Ranking × Rating
1	Low purchase price	S	10	£40 150	9	90
2	Good lifting capacities	T	10	Max 46.OT	7	70
3	Good dragline capacities	T	7	Max 6T	10	70
4	Good service facilities/ technical backing	S	7		7	49
5	Low running costs	S	8	No information	–	– –
6	Low upkeep cost/robust construction	S/T	4	4	7	28 28
7	Good safety features. Reliable safe load indicator	T	8	Weightload	10	80
8	Low machine weight/cost of transport between sites	S	2	55T	10	20
9	Low ground bearing pressure	T	3	9.06 psi (ave)	10	30
10	Reliable power unit/smooth operation	T	7	Rolls-Royce diesel	10	70
11	Fast slewing speed	T	5	3.18 rpm	10	50
12	Fast hoist speed (single line)	T	6	150 ft/min	10	60
13	Fast travel around site	T	4	0.750 mph	9	36
14	Familiarity with machine	S	4		6	24
15	Manoeuvrability/compact Geometry/travel on inclines	T	5	Max grad 1 in 4	10	50
16	Long working life/good trade-in price	S	8	9 years – £3 000	10	80
17	Short rigging time/ease of conversion to dragline	T	4		7	28
18	Operator view and comfort/ good layout of controls	T	5		7	35
S	Cost and service element	39%	43			291
T	Technical element	61%	68			607
	Totals	100%	111			898

Figure 7.2 (continued, opposite) Decision analysis sheet. Purpose of decision: evaluation of cranes for possible purchase and addition to company fleet

Alternatives				Adverse consequences			
HFC 416	Go/No	Sirac Highlift	Go/No	Harris 46T	P	I	P×I
Yes	✓	Yes	✓	Strike may		£	£
150′ max main	✓	160′ max main	✓	delay	0.3	1200	360
Yes	✓	Only released as lifting crane at present	X	delivery			
6/8 weeks	✓	8 weeks	✓				
Yes	✓	Yes	✓				
Yes	✓	Yes	✓				
Yes	✓	Yes	✓				

Information	Rat-ing	Weighted score	Information				
£39 482	10	100	£36 534		P	I	P×I
Max 53.6T	10	100	Max 45T				
Max 6T	8	56		HFC 416			
	10	70					
No information	–	–	No information	As a result of a design fault on the clutch some down-time may be necessary to effect a repair			
	10	40 40					
Wylie	9	72	Wylie				
66T	9	18	50T				
9.50 psi (ave)	9	27	8.50 psi (ave)			£	£
Dorman diesel	7	49	Dorman diesel		0.9	300	270
3.00 rpm	9	45	4.00 rpm				
140 ft/min	9	54	156 ft/min				
0.818 mph	10	40	0.907 mph				
	10	40					
Max grad 1 in 4	9	45	Max grad 1 in 4				
9 years – £3 000	10	80					
	10	40					
	10	50					
		348	This crane is removed from the analysis as *must* C is violated				
		618					
		966					

Step 4: Evaluate the Alternative against the Objectives

The stage is now reached where it is necessary to consider in detail the specification of the selected machines. Difficulties will now begin to emerge: for instance, there is rarely a unique price for a large item of construction equipment, as each manufacturer will offer features slightly different from those of its competitors' machines.

In this example the quotations received for the three cranes are shown in Table 7.1. It can be seen that the information assembled for each crane varies in its detail, and any decision on price can be only a compromise. However, within the limits imposed by these variations, the relative prices of the machines can be compared. But it must be remembered that discounts and payment arrangements may influence the final decision. Also, the cost and availability of spare parts and maintenance should not be overlooked at this stage, especially from foreign suppliers affected by exchange rate fluctuations.

Table 7.1

Component	Harris	HFC	Sirac
Base machine	£32 809	£26 833	£34 070
Cat head boom	(30ft) inc.	N/A	N/A
Taper top boom	(27ft) £756	(50ft) £3 915	inc.
Boom inserts to max.	£1 469	£2 524	inc.
Fly jib	(30ft) £573	(30ft) £561	(40ft) £1 482
Power lowering equipment	inc.	£2 135	inc.
Safe load indicator and tests	£1 830	£969	£902
Cab heater	inc.	£90	£80
Hook block	£213	£181	inc.
Parts to convert to dragline	£2 500	£968	N/A
$2\frac{1}{2}$ yd^3 dragline bucket	inc.	£1 306	N/A
Total cost	£40 150	£39 482	£36 534

Table 7.1 reveals that the Sirac Highlift was not quoted with dragline equipment and that, as it is a new model, it can only be used as a lift crane at present. Therefore, although this machine slipped through the preliminary screening of the *must* objectives, it is now marked *no go* against objective C in Figure 7.2 and is eliminated from further consideration. The Harris and HFC cranes comply with all the *must* objectives and effort can now be concentrated upon the *want* objectives of these two machines.

The *want* objectives are now separated into two groups – those which can be evaluated from the manufacturer's specification and those which cannot, as follows:

Can	Cannot
1. Purchase price	4. Maintenance requirements
2. Safe working loads	5. Running costs
3. Dragline range	15. Manoeuvrability
6. Method of assembly	17. Assembly time
7. Safety features	18. Driver comfort
8. Weight of machine	
9. Bearing pressure under tracks	
10. Power unit and size	
11. Slewing action	
12. Hoist system	
13. Travelling speed	

For those *wants* which are not readily available from manufacturers' information, more subtle sources must be found – for example, site demonstrations, visits to the manufacturer, records of data on experiences with similar machines, personnel contacts with other plant users and discussions with machine operators.

With sufficient information on all the *wants* objectives available, the two remaining crane alternatives (namely Harris 46T and HFC 416) are compared for performance. Each *want* objective is considered in turn and rated on a 0–10 scale. The technical element and economic element ratings are separated in order that the importance of each can be ascertained.

To obtain the relative worth of each *want* objective, the *rating* value is multiplied by the previously given *rank* number. The resulting weighted score represents the performance of the crane against its objective.

The weighted scores for each objective are subsequently added, to give totals for each crane alternative, and the totals for each provide an indication of the relative position with respect to the specified objectives. In this example the total weighted scores shown in Figure 7.2 are summarized as follows:

	Percentage of total	*Harris*	*HFC*
Cost and service element (S)	39	291	348
Technical element (T)	61	607	618
	100	898	966

Note: It must be emphasised that these values are determined largely by subjective judgement, albeit based on careful analysis of the available facts, and as such do not make the selection of the best alternative a routine procedure. However, the values do make it possible to deal systematically with many factors which otherwise could not easily be related for importance.

Step 5: Choose the Best Alternative as the Preliminary Decision

The alternative that receives the highest weighted score is presumably the best option — in this case HFC 416. It is, of course, not a perfect choice — for instance, one would probably prefer a machine with a Rolls-Royce engine. But the choice is at least one which strikes a reasonable balance between the good and bad features of the machine. Fortunately, both the technical and cost element totals are greater for the HFC 416. The decision would be far more difficult if one of the elements were larger for the Harris 46T crane. Some further balance would then be needed between economic resources and the technical benefits accruing to the company.

The preliminary choice is the one which best satisfies the objectives overall. It is thus a compromise, as undoubtedly the alternatives have some superior features. But the method used shows the manager how he arrived at this decision and therefore where the possible pitfalls may lie.

Step 6: Re-evaluate the Decision and Assess the Adverse Possibilities of That Choice

When many alternatives are available, the weighted scores of one or two are sometimes quite close, so, before the final decision is made, any adverse consequences must be considered. The manager must look for snags, potential shortcomings or anything else that could go wrong. The probability (P) of the adverse consequences should be assessed and a seriousness weighting (I) given to its possible effect. An expression of the total degree of threat may be obtained by multiplying the seriousness weighting by the probability estimate.

In this example the plant and site managers arrange a meeting to discuss the implications of the proposals, and the following points are raised:

(1) The discarded Sirac Highlift appeared to be a fairly good crane on preliminary inspection and perhaps the *must* objective necessitating dragline conversion could have been reduced to a major *want* objective. A further check on the records reveals that only 20% of the time was spent on such duties. Thus, unless the plant manager insists that this facility is essential, this crane should be reconsidered. But the *must* objective should only be waived in very exceptional circumstances.

(2) Possible adverse consequences of choosing the Harris 46T, such as the effects of a strike at the factory, can be assessed from past

experience. It may be that on the past ten occasions when a new wage agreement has been under negotiation between the Harris company and its workforce, a strike resulted on three occasions. So the probability (P) of a strike at this time is judged to be 30%. If a strike did occur which would last more than a few weeks, then the delivery of this machine would exceed the specified time on the order, in which case the crane should not be considered at all. However, should a strike occur lasting 4 weeks, the cost of hiring a crane would be about £300 per week; therefore, the degree of threat is $(4 \times 300) \times 0.3 = £360$.

(3) One possible adverse consequence of the HFC 416 has come to light. A serious accident has just revealed a major design fault in the clutch mechanism of the winch. A temporary modification will be made to all existing models and stocks held. But a completely new part is not available for 6 months and will require taking the crane out of service for 1 week at that time. The probability that this service is required is 90% and the cost of hire for another crane is £300 per week, thereby directly increasing the contract cost at this rate. The degree of threat is $(1 \times 300) \times 0.9 = £270$. The choice, therefore, is still the HFC 416. Had the consequences been reversed, a much closer examination of the advantages and disadvantages would be necessary, making the decision much more dificult.

Step 7: Set up Contingency Plans to Control the Effects of the Final Decision

The adverse consequences represent potential problems, and they must be prevented from causing too much inconvenience. This is done either by taking preventative action to remove the cause or by deciding upon a contingency action if the potential problem occurs. In our case, the simple remedy is to ensure that alternative machines are available for hire at suitable times or to arrange work on site so that interruption is minimal.

Comments

A decision has been reached to buy the HFC 416, and this is the best choice on the basis of the judgement of the plant manager concerned. For someone else with different experience, the assessment of the objectives might be quite different and would possibly produce a different result. But even accepting this obvious weakness, the method at least forces the manager to consider most of the facts. It provides a record of the thought processes and will help to sort out points of confusion, for it is almost impossible to memorise all the facts and relationships, except for the very simple choices. In such cases the decision process can be adequately carried out mentally without the need for the elaborate method described. Finally, the decision process is written down on paper and is available for all to see.

The whole procedure can now be readily performed on a personal computer, using commercial spreadsheet and database packages. Indeed, equipment manufacturers themselves may eventually offer comprehensive information such as machine specifications, performance characteristics, prices, etc., on computer discs, which would considerably assist the analysis process.

BIBLIOGRAPHY

Anthill, S. M., Ryan, P. W. S. and Easton, G. R. (1989). *Civil Engineering Construction*, McGraw-Hill

Dixon, J. R. (1966). *Design Engineering – Inventiveness Analysis and Decision Making*, McGraw-Hill

Harris, F. (1983). *Ground Engineering Equipment and Methods*, BSP Professional

Harris, F. (1989). *Modern Construction Equipment and Methods*, Longman

Kepner, C. H. and Tregoe, B. B. (1975). *The Rational Manager*, McGraw-Hill

Nunnally, S. W. (1985). *Managing Construction Equipment*, Prentice-Hall

Peurifoy, R. L. and Ledbetter, W. B. (1985). *Construction Planning, Equipment and Methods*, McGraw-Hill

Russell, J. E. (1985). *Construction Equipment*, Reston

Chapter 8
Calculating a Hire Rate

8.1 INTRODUCTION

Revenues from equipment hired out or rented to the market or owned and operated for internal use by a construction company should not only recover the owning and operating costs, but also achieve an additional profit.

Ownership costs are fixed costs arising indirectly, such as company overheads. Such costs are incurred throughout the period of ownership and are a fixed charge. They include: (1) cost of capital; (2) depreciation on the equipment; (3) insurances and licences; (4) corporation tax and capital allowances; (5) establishment charges.

The operating costs are direct costs of material, labour and expenses, which vary according to the usage of the machine, and include: (1) servicing costs — e.g. oil, grease and other consumables; (2) maintenance costs; (3) transport charges; (4) fuel; (5) operators' wages.

8.2 COST OF CAPITAL

Capital to purchase the machine or item of equipment is generally obtained from borrowing or retained profits. In either case the interest charge on the loan or an acceptable rate of return on private funds should be incorporated into the hire rate.

8.3 DEPRECIATION

Most items of equipment wear out and deteriorate with usage or become obsolete with time. Thus, at the end of its economic life sufficient revenues should have been generated to cancel out the capital part of a loan or to replace the machine when internal funds are used. In calculating depreciation, it is necessary to estimate the resale or scrap value of the machine and the useful life.

Estimation of the useful life is mostly done by intuition based on experience of operating a similar item — e.g. when the utilisation level shows a clear downward trend accompanied by increased maintenance costs and loss of profit, the machine has probably reached the end of its useful life for the company.

There are several methods of calculating the annual depreciation charge, but the choice will depend largely upon the type of machine and its operation. In principle, depreciation is deducted from profits (since it represents a cost on the business activities) before tax is paid. It is, therefore, necessary to obtain approval of the method and depreciation period from the Inspector of Taxes before a suitable policy can be adopted. However, where capital allowances are in operation special arrangements generally apply (see Chapters 5 and 14).

8.3.1   Straight Line Depreciation

EXAMPLE 8.1

It is decided to purchase a mechanical excavator costing £42 000 to work on average 2 000 hours per year. The life of the machine is expected to be 10 years, after which time the salvage value will be £2 000.

Purchase price	=	£42 000
Residual value	=	£ 2 000
∴ Total depreciation	=	£40 000
∴ Annual depreciation	=	£ 4 000
∴ Hourly depreciation	=	£ 2

8.3.2   Declining Balance Depreciation

Table 8.1   Declining balance depreciation example (at 26.2%)

End of year	Depreciation (£)	Depreciation for year (£)	Book value (£)
0	26.2	0	42 000
1	26.2	11 004	30 996
2	26.2	8 121	22 875
3	26.2	5 994	16 881
4	26.2	4 423	12 458
5	26.2	3 264	9 194
6	26.2	2 409	6 785
7	26.2	1 778	5 007
8	26.2	1 312	3 695
9	26.2	968	2 727
10	26.2	727	2 000

Total: £40 000

Using the example on page 102, suppose that this time the depreciation is made on a fixed percentage basis rather than a fixed sum.

The figures in Table 8.1 can more easily be calculated from the formula:

$$d = (1 - n\sqrt{L/P}) \times 100$$

where L = salvage value, P = purchase price, n = life of asset and d = percentage depreciation.

8.3.3   Sinking Fund Method of Depreciation

A fixed sum is put aside from revenue each year and invested with compound interest throughout the life of the asset. After successive instalments the sum accumulates, to produce the original purchase price less the scrap value.

Taking the given example and assuming that 6% interest is earned on savings, the annual amount required is £3 034, thus:

Uniform series factor that amounts to 1 over 10 years = 0.075 86
Therefore, annual payment = 0.075 86 × (42 000 − 2 000 = £3 034

Table 8.2 gives a detailed breakdown of the analysis.

Table 8.2 Sinking fund example (6% interest on savings)

Year	Payment (£)	Interest (£)	Depreciation (£)	Book value (£)
1	3 034	0	3 034	38 966
2	3 034	182	3 216	35 750
3	3 034	375	3 409	32 341
4	3 034	581	3 615	28 726
5	3 034	798	3 832	24 894
6	3 034	1 028	4 062	20 832
7	3 034	1 271	4 305	16 527
8	3 034	1 529	4 563	11 964
9	3 034	1 803	4 837	7 127
10	3 034	2 093	5 127	2 000

Total: £40 000

8.3.4   Sum of Digits Method

Life   = 10 years
Digits = 1 + 2 + 3 + 4 + 5 + 6 + 7 + 8 + 9 + 10 = 55

Table 8.3 Sum of digits depreciation example

Year	Factor	Depreciation (£)	Book value (£)
1	10/55	7 273	34 727
2	9/55	6 545	28 182
3	8/55	5 818	22 364
4	7/55	5 091	17 273
5	6/55	4 364	12 909
6	5/55	3 636	9 273
7	4/55	2 909	6 364
8	3/55	2 182	4 182
9	2/55	1 454	2 728
10	1/55	728	2 000

Total: £40 000

8.3.5   Free Depreciation

The asset is totally depreciated initially by this method. For example, the item of equipment purchased for £42 000 would be depreciated by £40 000 immediately on acquisition, leaving only the £2 000 salvage value.

Graphical Comparison of the Depreciation Methods

Figure 8.1 demonstrates the main features of the declining balance method and the straight line method. In the early years the asset is heavily written down, which is particularly helpful in the case of construction plant, as the repair and maintenance costs are likely to be low when new, but increase with age and usage.

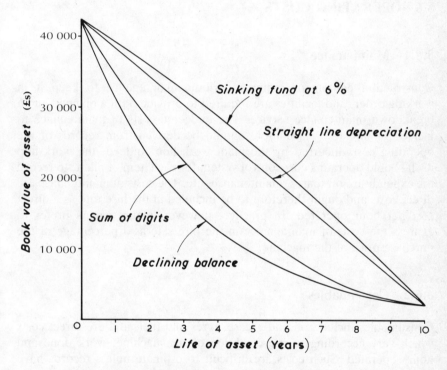

Figure 8.1 Graphical comparison of depreciation methods

8.4 LICENCES AND INSURANCES

The costs for licences and insurance depend upon whether the equipment is to be used on the public roads or otherwise. In general, the amount of insurance cover recommended depends upon the degree of risk to third parties, adjacent property or public utilities. These aspects are fully discussed in Chapter 11.

8.5   ESTABLISHMENT CHARGES

The company must recover the cost of its overheads in the hire rate. Overheads relate to the central organisation, and include the offices and workshops and the associated administrative facilities. These fixed costs may be apportioned on the basis of budgeted annual plant operating hours or as a percentage of the purchase price of the machine per annum.

8.6   OPERATING COSTS

8.6.1   Maintenance

Construction equipment requires periodic maintenance to keep it in working order, and facilities are required to provide both a planned and a breakdown maintenance service. The budgeted costs of maintenance to include in hire rate calculations should be derived from records of the operating costs incurred by the plant workshop. Indeed, the workshop itself should operate a cost control system for each item of plant, to record the expenditure on spares, maintenance, etc. The costs of maintenance are direct costs and ought therefore to be included in the hire rate as a direct cost per hour operated. In practice, however, many owners prefer to express the costs of maintenance in the hire rate as a percentage of the purchase price of the machine.

8.6.2   Consumables

Consumables include oil and grease, tyres and fuel, and are direct costs which vary according to the condition of the machine, work done and hours operated. Such costs are difficult to estimate unless records have been kept from operating similar equipment in the past. However, manufacturers provide guidelines on the consumption of these materials, but care should be exercised, as the data are likely to relate to new equipment operating under ideal conditions.

The inclusion of the cost of fuel in the hire rate will, of course, depend upon the hire contract conditions.

8.6.3   Operators' Wages

Operators' wages are not usually included in the hire rate, as many items are hired exclusive of the operator. But, when required, costs based on the driver's hourly rate must be recovered, with allowances for overtime, bonus, travelling subsistence, national insurance premiums, holiday pay, sick pay, pensions, etc.

8.7   THE EFFECT OF CORPORATION TAX AND CAPITAL ALLOWANCES

These are treated more fully in Chapters 5 and 14 but, briefly, depending upon national policy, governments may introduce investment incentives to encourage companies to account for free depreciation on equipment, so that an item is allowed the full purchase price, to be deducted from company profits in the year of purchase, and thereby the amount of corporation tax due is reduced in that year. The tax, however, will, of course, be recovered in full from profits in subsequent years, since there is no allowance to set aside thereafter. The effect of this arrangement is to alter the sequence of cash flows for the business, which may produce a more favourable return on capital employed and so encourage other investment.

It is emphasised that capital allowances are basically a feature of the profit and loss account for the whole enterprise for the calculation of corporation tax payments. Clearly, to allow full initial depreciation in the hire rate would be nonsensical, as the rate would be uncompetitive during that year. Thus, internal depreciation should be based on a realistic assessment of the life of the asset.

The effect of the 100% capital allowance as operated until recently in the UK was to encourage firms to purchase equipment to set against profits as a popular tax avoidance device. Consequently, the market became grossly oversupplied with plant for hire with the unfortunate result that hire rates fell below levels which could earn a healthy rate of return on capital employed. Thus, although over the short term customers and both home and overseas manufacturers benefited, the plant hire sector itself could not sustain low profitability indefinitely and considerable adjustment has taken place through bankruptcy and mergers.

8.8   CALCULATING THE ECONOMIC HIRE RATE

There are several acceptable methods of calculating an economic hire rate. The most favoured is the simple calculation to allow for ownership and operating costs with a contribution for profit. However, a more satisfactory method for investment extending over a few years is the discounted cash flow yield (DCF), which takes into account the timing of cash flows.

EXAMPLE 8.2: THE CONVENTIONAL METHOD

An excavator, crawler-mounted, of $1\frac{1}{2}$ m^3 capacity, is purchased new for £46 000. Its estimated life is 10 years, with a resale value of £4 000.

Capital cost	£46 000
Resale value	£4 000
Anticipated life	10 years
Insurance premium	£200 per year
Road tax and licences	£100 per year
Maintenance	10% of capital cost
Consumables	£400 per year
Overheads of business	£2 per hour
Required rate of return on investment	15% per year
Budgeted operating time	2 000 hours per year
Transport charges	say £100

Item	£ per year
Depreciation (straight line) = $\dfrac{42\,000}{10}$	4 200
Interest on finance, calculated using a capital recovery factor from interest tables (CRF = 0.199 at 15% p.a. for 10 years): $\dfrac{(46\,000 \times 0.199 \times 10) - 46\,000}{10}$	4 554
Fixed overheads = 2 × 2000	4 000
Road tax and licences	100
Insurance premium	200
Ownership (fixed) cost	13 054

Item	£ per year
Consumables	400
Maintenance × 10% × 46 000	4 600
Operating cost (variable)	5 000
Total cost	18 054

Hire charge = 18 054/2 000 = £9.03 per hour (or £361 per 40 hour week).

The cost of transport and any additional profit should be added to this figure.

Finally, the hire rate is based on a utilisation period of 2 000 hours, and should this target not be reached, the ownership (fixed) costs will be underrecovered and budgeted profit will not be achieved.

8.8.1   Alternative Analysis Using DCF (See Chapter 5)

DCF takes into account the timing of cash flows, whereby income and outgoings are balanced to yield a satisfactory return. The problem is thus restructured as shown in Table 8.4.

Information:

(1) 4.771 5 is the net present worth factor of 1 per period for 9 years at 15% interest rate.
(2) 0.247 2 is the net present worth factor of 1 at year 10 for a 15% interest rate.

To return 15% on the investment over 10 years, the total cash flows reduced to net present worth at time zero must equate to:

$$0 = -50\ 300 + (x - 9\ 300) \times 4.771\ 5 + (x - 1\ 000) \times 0.247\ 18)$$
$$0 = -51\ 300 + 4.771\ 5x - 25\ 289 + 0.247\ 2x - 247$$
$$5.02x = 94\ 922$$
$$x = 18\ 913$$
$$\therefore \text{Hire charge} = 18\ 913/2\ 000 = £9.45 \text{ per hour}$$

It can be seen that the DCF method automatically accounts for the depreciation over the life of the asset.

8.8.2   The Effect of Inflation

The two methods outlined above have not considered the effects of inflation on the value of the investment, the consequences of which are emphasised by the following example.

Table 8.4 DCF analysis of a hire rate

Year	A Capital cost (£)	B Resale value (£)	C Operating costs (£)	D Ownership costs (£)	A + B + C + D = E Cash out (£)	Revenue, F Cash in (£)	E + F Total (£)
0	− 46 000		− 0	− 4 300	− 50 300	0	− 50 300
1			− 5 000	− 4 300	− 9 300	x	x −
2			− 5 000	− 4 300	− 9 300	x	x −
3			− 5 000	− 4 300	− 9 300	x	x −
4			− 5 000	− 4 300	− 9 300	x	x −
5			− 5 000	− 4 300	− 9 300	x	x −
6			− 5 000	− 4 300	− 9 300	x	x −
7			− 5 000	− 4 300	− 9 300	x	x −
8			− 5 000	− 4 300	− 9 300	x	x −
9			− 5 000	− 4 300	− 9 300	x	x −
10		+ 4 000	− 5 000	0	+ 1 000	x	x − 1 000

EXAMPLE 8.3

Purchase price	£46 000
Resale value after 10 years	£ 4 000
	£42 000

Annual depreciation $\dfrac{£42\ 000}{10} = £\ \ 4\ 200$

Inflation at 10% per year

Purchase price after 10 years	£4 600 × 2.60 = £119 646
Resale value after 10 years	£4 000 × 2.60 = £ 10 400
	£109 246

Average annual depreciation   = £109 246/10 = £10 924
The shortfall = £67 246

Depreciation is the most vulnerable element in the hire rate calculation, because other items are the result of obvious cost movements — e.g. materials, wages, interest rates. The hire rate should therefore be revised frequently, to keep up with inflation. In fact, it may even be necessary to provide for 'backlog' depreciation to allow for the period prior to an underestimated price rise. Assessments can be made from the Baxter special price indices for construction plant, or, alternatively, the small operator replacement prices should be obtained from plant dealers and manufacturers.

Inflation produces other severe effects by causing trading profits to be overstated. As a result corporation tax is paid on the inflated rather than the real profit.

Adjustment of a Hire Rate for Inflation

If the capital to purchase an item of equipment is either borrowed or alternatively required by hire purchase, the hire rate would simply be adjusted in line with the terms imposed on the loan, as the total capital sum remains fixed irrespective of the level of inflation. Only the level of interest may fluctuate. However, when the item is to be purchased from internal funds, the value of the asset must be periodically revised to keep up with inflation, if the consequences of a shortfall in depreciation provision and in real terms a negative rate of return on capital are to be avoided. The effect of using a 10% inflation price index on the original example is shown below.

EXAMPLE 8.4

Inflation is 10% per annum over 10 years. The purchase price is £46 000 and the historical resale value is £4 000.

Year	Index	Replacement price (£)	Accumulated historical depreciation (£)	Accumulated inflated depreciation (£)	Book value (£)
0	100.0	46 000	0	0	46 000
1	110.0	50 600	4 200	4 620	45 980
2	121.0	55 660	8 400	10 164	45 496
3	133.1	61 226	12 600	16 771	44 455
4	146.4	67 344	16 800	24 595	42 749
5	160.7	73 922	21 000	33 749	40 173
6	176.8	81 328	25 200	44 554	36 774
7	194.5	89 470	29 400	57 183	32 287
8	214.9	98 854	33 600	77 206	21 648
9	236.4	108 744	37 800	89 359	19 385
10	260.1	119 646	42 000	109 242	10 400

The depreciation in, say, the first year of inflation is £4 620.

Interest on Finance

With inflation at 10% the apparent rate of return must be used in the calculations:

$$(1 + i_a) = (1 + i_r)(1 + i_d)$$

where i_a = apparent rate of return; i_r = real rate of return; i_d = rate of inflation. Therefore,

$$(1 + i_a) = (1 + 0.15)(1 + 0.1) = 1.265$$
$$i_a = 0.265 = 26.5\%$$

Thus, interest on finance, using a capital recovery factor of 26.5% from interest tables,

$$= \frac{(46\ 000 \times 0.292\ 9 \times 10) - 46\ 000}{10} = £8\ 873$$

Other Items

Fixed overhead	£4 000
Road tax and licences	£ 100
Insurance premium	£ 200
Consumables	£ 400
Maintenance	£4 600
	£9 300

Multiplying the total figure by the first year index
$= (9\ 300 \times 110)/100 = £10\ 230.$
Therefore, the hire charge $= £\ 4\ 620$
$£\ 8\ 873$
$£10\ 230$

$£23\ 723 \div 2\ 000 = £11.86$ per hour (or
£474.46 per week)

If inflation continued as shown by the indices, the hire rate for year 5 should be given by:

Depreciation for year 5 $= 33\ 749 - 24\ 595 = £\ 9\ 154$
Interest on finance $= £\ 8\ 873$
Other items $= 9\ 300 \times (160.7/100)$ $= £15\ 531$

$£33\ 558,$

i.e. £16.78 per hour

It can be seen that during periods of inflation the hire rate should be revised at least annually. When inflation exceeds about 10%, the revision may need to be at quarterly intervals, depending upon the demands for payment by suppliers of materials, etc. However, this is not always possible when the market for hired equipment is slack and competitors are willing to undercut hire rates.

When inflation exceeds the net rate of interest on borrowed capital (i.e. interest on borrowings is deducted from company profits before corporation tax is paid, which in effect produces a lower rate of interest), a 3 year replacement cycle for plant is best. Below that rate of inflation a longer cycle is preferable. The cross-over point in the choice of the method of financing is when inflation is equal to the net of tax interest rate on capital. Below this rate of inflation self-finance is more attractive than borrowed money, and vice versa.

8.8.3 Method of Calculating Hire Rates Adopted in Practice

Small Companies

Small companies tend to set the hire rate in accordance with the market levels, often guided by the figures published regularly by the CPA, HAE, etc.

Indeed, most users are well aware of the prevailing market rate for a machine and each hire is generally negotiated at around this figure. The experienced hirer is usually well informed of the availability of machines locally and thus able to negotiate very competitively. The minimum hire rate, especially for the owner-operator, often relates to payments needed for repayment of the loan on the piece of equipment plus a sum to cover running costs and the operators' overheads, the latter being perhaps as minimal as providing the owner's salary.

Medium-to-large Companies

Large companies with comprehensive central and regional plant depots carry out detailed analyses of the economics of owning and operating equipment. The hire rates are usually based on collected cost data of the firm's operations and calculated by the principles established above, including DCF techniques.

8.9 THE EFFECTS OF ECONOMIC RECESSION ON PLANT HIRE

During the period following the oil crisis of the 1970s, the volume of construction work steadily declined, with an attendant reduction in the need for construction plant. Faced with a declining demand, many owners of equipment cut back the size of their fleets in order to survive. However, earlier good maintenance ensured acceptable resale values to overseas markets and, fortunately, many machines which were retained were purchased in the days of low interest rates. Therefore, firms continued to hire out plant at competitive rates, all aided by government incentives and a wide choice of manufacturers offering generous cash discounts of up to 20% on new purchases. As a consequence, machines were 'run into the ground' as the rates of hire became insufficient to cover adequate maintenance and servicing. Some restructuring of the market for plant hire subsequently took place through mergers and concentration on specialised needs in the construction industry for the medium-to-large plant owner, leaving the more general market for the small competitive local enterprise. However, the upturn in the economy during the 1980s has tended to mask the inherent instability of the market, and if recession returns, undoubtedly further restructuring will be inevitable.

BIBLIOGRAPHY

Caterpillar Company, Peoria, Illinois (1989). *Equipment Economics*, Market Development Division

Construction Plant-Hire Association. *CPA Handbook*, (annual)

Douglas, J. (1978). *Construction Equipment Policy*, McGraw-Hill

Eagleston, F. N. (1983). *Insurance for the Construction Industry*, Longman

General Motors Corporation (1989). *Production and Cost Estimating of Material Movement*, Terex Division

Harris, F. and McCaffer, R. (1986). *Worked Examples in Construction Management*, 2nd edn, BSP Professional

Harris, F. and McCaffer, R. (1989). *Modern Construction Management*, 3rd edn, BSP Professional

International Harvester Company of Great Britain (1989). *Basic Estimating*

Komatsu Ltd (1989). *Specification and Applications Handbook*

McCaffer, R. and Baldwin, A. (1985). *Estimating and Tendering for Civil Engineering Works*, BSP Professional

Mead, H. T. and Mitchell, G. L. (1972). *Plant Hire for Building and Construction*, Newnes-Butterworths

Peurifoy, R. L. and Ledbetter, W. B. (1985). *Construction Planning, Equipment and Methods*, McGraw-Hill

Powell-Smith, V. (1989). *The Model Conditions for the Hiring of Plant*, CPA

Section 3

Operational Management

Chapter 9
Maintenance of Equipment

9.1 PLANT MAINTENANCE OBJECTIVES

Construction equipment, like any other machine, can be expected to break down during its working life. This may be due to normal wear and tear, or a sudden failure of a component part.

The primary purpose of providing maintenance is to reduce the incidence of failure, by either replacement, repair or servicing, in order to achieve an economical level of utilisation during the working life of the machine. A reduction in 'down-time' will minimise costly stoppages on the construction site and the disruptive effect on labour and the programme of work. However, while the purpose of maintenance is to keep the equipment in service, this must not be achieved at the expense of safety. The costs of maintenance must be balanced against the benefits, and at some stage a plant item will require complete replacement by a new machine.

9.2 PLANT MAINTENANCE POLICY

The alternative maintenance policy options are shown in Figure 9.1. The extent to which a company applies one or other of the alternatives will depend upon the type of holding and the firm's attitude to maintenance in general and the safety regulations. However, a policy directed towards ensuring controlled, regular and disciplined maintenance procedures is essential if basic maintenance objectives are to succeed. In theory, therefore, planned maintenance should be the backbone of the system, but the practical difficulties arising from the dispersed nature of construction work generally require aspects of planned, unplanned and replacement maintenance policies to be combined.

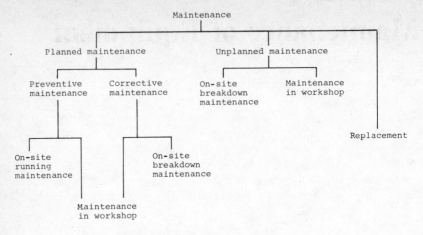

Figure 9.1   Maintenance options

9.2.1   Planned Preventive Maintenance

This system requires the implementation of planned regular procedures directed towards ensuring the efficient use of equipment by reducing the incidence of breakdowns. The maintenance actions required are:

(1) Daily servicing and superficial inspection. This aspect is often adequately achieved by allowing the machine operator half an hour before and after normal working hours to carry out the tasks.
(2) Regular full maintenance and inspection, including periodic overhaul.
(3) Replacement or repair of component parts within a working period based on the expected duties and conditions.

Clearly, this is a very comprehensive service and can be operated only by those firms with extensive holdings, where 'down-time' can be avoided by substituting other machines and where workshop facilities are available.

9.2.2   Planned Corrective Maintenance

Adequate maintenance is performed to enable a machine to operate while on site, but a major overhaul should be undertaken after its duties are completed. The procedures are, therefore, slightly less systematic compared with preventive maintenance, as components are usually replaced only during the full inspection and servicing period or when a breakdown occurs or is anticipated. This policy does not ensure that the highest safety standards are upheld, but is often favoured by construction companies which hold few items of plant, where full maintenance resources and personnel are available only at central workshops.

9.2.3   Unplanned Maintenance

This system is adopted when the costs of regular maintenance are likely to exceed the cost of complete replacement or where it is not economically justifiable to carry out maintenance until either the machine breaks down or the operating efficiency becomes unacceptable. Clearly, this would be appropriate only for equipment which is not essential to the production process and the failure of which would neither cause considerable disruption nor constitute a safety hazard. As a consequence, unplanned maintenance is rarely appropriate for construction machinery, except for small tools, punctures, etc., although in practice all too many firms adopt this policy for major equipment, with disastrous consequences for production efficiency.

9.2.4   Replacement

Most items of construction equipment have a life exceeding the point at which a major overhaul is required, and the question of replacement should therefore arise only when the costs of maintenance exceed the benefits of operating the item. However, the state of the second-hand market can fluctuate over the short term to provide a profitable opportunity for selling a machine before its planned replacement period. Alternatively, superior equipment may become available, to outweigh the advantages of holding outdated machinery.

9.2.5   The Merits of Planned Maintenance

These are:

(1) Improved utilisation levels of equipment.
(2) Maintenance periods can be co-ordinated with site production requirements.
(3) Spare parts can be obtained in good time and stock maintained at adequate levels.
(4) Regular work schedules for maintenance personnel can be programmed facilitating the allocation of fitters and mechanics specifically where specialist skills and experience are required.

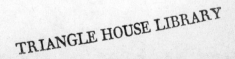

(5) The overall result enhances the awareness and importance of maintenance in the well-being of the company.

Planned maintenance is expensive and is used to maximum advantage where the effects of breakdowns would be extremely damaging. The opportunities for carrying out effective inspection and overhaul for many items of construction equipment often occurs during the transfer from one site to the next and a policy of corrective rather than preventive maintenance is sometimes preferred. Both systems require a comprehensive maintenance organisation, but the costs can be recovered from the improved working efficiency of the plant, unlike the alternative unplanned arrangement.

9.2.6   Factors Affecting Maintenance Policy

Planned maintenance offers the most reliable policy, but involves the setting up of workshops, offices and stores, coupled with a significant investment in tools and trained staff. Also, the operational problems imposed by the nature of construction work severely restrict the quality and amount of maintenance provision which may be achieved.

In a large national construction company a centralised maintenance facility is often too expensive to operate for servicing plant items spread over a wide location. The alternative of establishing workshop facilities on individual sites may be justified on the larger and/or more remote contracts, but is generally prohibitively expensive on the typically small site. Regionalisation offers a suitable compromise, whereby individual maintenance facilities may be set up on the large, equipment-intensive sites, perhaps co-ordinated from the regional depot, with a mobile workshop operating from the regional facility to serve the smaller sites. This system is popular with plant hire companies and is being adopted increasingly by construction companies. The mobile workshop can be equipped to carry out major servicing and repairs, using a vehicle such as a Land Rover to cope with most of the adverse conditions found on construction sites. A complete overhaul may be performed later at the regional depot.

9.2.7   Manufacturer's Contribution

The assistance offered to equipment owners by individual manufacturers or their agents varies. Evidence suggests that some manufacturers view after-sales service as being entirely the responsibility of their appointed distributor. Others consider maintenance, in particular, to be a major consideration and provide extensive services both directly and through distributors.

The services provided generally fall within the following alternatives:

(1) The manufacturer or distributor of the equipment provides a full maintenance service as part of the purchase contract.
(2) The manufacturer or distributor makes available a back-up service in respect of field inspection, field component replacements and workshop overhauls — e.g. for more specialised maintenance needs such as transmissions and engines.
(3) Provision of spare parts only by the manufacturer.

The more efficient manufacturer or distributor would tend to have the capability to provide the first alternative and would often include the following facilities:

(1) A nationwide distribution network capable of undertaking maintenance, repairs and overhauls, and stocking a full complement of spare parts.
(2) Technical support facilities, including mobile breakdown units.
(3) Training programmes designed for distributor's and client's operators and maintenance personnel, including a mobile unit which can visit customers to provide training courses.
(4) Fully detailed and planned preventive maintenance programme for each type of machine.
(5) Firm guarantees concerning the maintenance commitment required from distributors.
(6) Supply by the manufacturer of reconditioned parts to reduce major overhaul costs.
(7) Provision of specially designed tools to aid maintenance work.
(8) Standardisation of components in different machines to reduce the stock level of spare parts.

9.3   STRATEGY FOR PLANT MAINTENANCE

The essential aim of all the maintenance policies described is to keep equipment in working order and so increase its productivity. However, the strategy required to achieve this objective demands the implementation of technical and administrative procedures which inevitably incur costs, and for any organisation, depending upon its maintenance efficiency, there is an optimum level of maintenance provision, as shown in Figure 9.2. This implies that at some level the cost of providing the maintenance service will exceed the costs of machine 'down-time'. Thus, not only is it important to

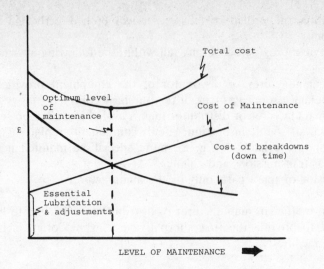

Figure 9.2 Optimum maintenance provision

install the correct maintenance procedures, but also the costs must be maintained and controlled. These can be considered in two classes — direct costs and indirect costs.

9.3.1   Direct Maintenance Costs

The first step is to prepare a maintenance budget based on the needs of the equipment holding and any items to be added during the life of the budget. Reference should be made to historical records of:

- Breakdown maintenance labour costs.
- Planned maintenance labour costs.
- Materials costs and fuel consumption.
- Spares costs.
- Administrative, technical, equipment and other overhead costs.

The budget provides the basis for monitoring the trend in overall maintenance effectiveness. In order to collect costing information, each item is identified with a cost centre, which might be a single important piece of equipment or alternatively a group of similar machines. The extent and depth of the information will vary between firms but, as a basis for information flow, a cost and maintenance control system should contain the following features:

(1) A *register of all equipment assets*, detailing the type, classification, purchase price, location, life, age, value and condition of each.

(2) A schedule of the exact *maintenance tasks* required on each item on a routine basis, together with the extent, details and methods to be employed.

(3) A *programme of events* defining the frequency of these maintenance tasks.

(4) An effective *history record* for each item, to ensure that the maintenance has been performed on schedule and in the correct manner. This will include a weekly return of hours operated or miles travelled and fuel consumed, coupled with a record of maintenance work carried out and the cost, including the materials used and spare parts supplied. The history record card is updated regularly with maintenance information each time a job report card has been completed.

(5) A cost recording system to facilitate monitoring the effectiveness of the maintenance effort measured against the budget.

The effectiveness of this procedure depends upon a disciplined application of the checking system and should ideally follow the stages indicated in Figure 9.3.

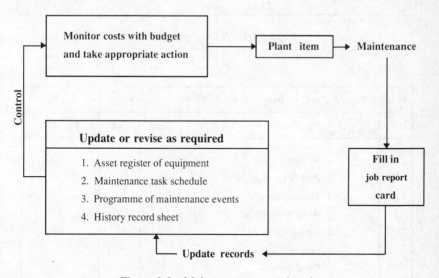

Figure 9.3   Maintenance control cycle

Job Report Card

The job report card should be completed by the mechanic after work has been carried out, and should contain a statement of the work done, the materials used and the condition of the unit. Subsequently, this information is used to update the history record card. See Figures 9.4 and 9.5.

Job Report Card No.	Date
Name & Details of Mechanic	
Equipment item	**Location**
Defects **Action taken** **Spares/Materials used** **Condition/Observations** **Remarks** **Time taken** **Remaining defects**	

Figure 9.4   A job report card

History Record Card		Date From To		Page No.		
Date	Job Report No.	Summary of Details	Time	Cost	Remarks
		Hours operated Fuel & Lubricants used Maintenance or Breakdown Cause Action taken Spares/Materials Condition			
Equipment item					
Location					

Figure 9.5   A history record card

9.3.2   Indirect Costs

When equipment breaks down, there is a loss of production while waiting for repair and during the repair itself. For the hire firm this may cause a reduction of the budgeted utilisation period and a loss in revenue, but for an owner-contractor the idle time of the construction work force must also be considered. It is important to consider these costs when preparing an overall budget for owning and operating the machine, including the cost of maintenance. Clearly, it is essential to decide upon realistic levels of utilisation time over, say, a yearly budget.

9.3.3   Safety Inspections

The various safety regulations (see Chapters 10 and 11) require the inspection and testing of equipment. All lifting tackle and most other items of equipment, when first acquired and before use, should be tested and examined by a competent person who should certify that the equipment is suitable for use and issue a certificate stating so. In most cases this certificate will be provided by the manufacturers but, if not, an insurance engineer can issue such a certificate. Once the item is in use the regulations also call for periodic testing by a competent person, the time interval depending upon the type of equipment, and a certificate issued by that person.

  Because the various regulations stipulate different inspection periods, the recommended procedures to adopt should consist of a combination of self-inspection by the owner or hirer at say three-monthly intervals, and then externally examined within the legally stated testing period. This latter inspection could be performed by a qualified engineer from the engineering inspection department of an insurance company. The services of the insurance company, however, do not relieve the owner of the equipment from the legal responsibility for periodic examination. Although most insurance companies have their own methods to try to ensure that the statutory periods are met, they are not legally bound to do this. Also the insurance engineer, having arrived at the premises will not necessarily search for equipment to test or get it ready for examination, and if the items are not to hand, then the inspection will most likely be missed. Clearly, therefore, it is in the interests of the equipment owner to install proper control procedures for inspections.

Control Procedures for Testing and Examination

Whenever possible, responsibility for arranging testing and examination should be given to a single individual. In this context, the asset register can assist in the time-tabling of inspections for major equipment, but for small items, such as lifting tackle, a physical audit may be required, since individual employees and machine operators often hoard items like slings etc. in tool boxes, lockers and cabs. To minimise this location problem, all equipment should be issued only through the stores under the control of a storekeeper and records maintained at least to the standard demanded by the legal statutes. The law requires that records be kept of the following details for each item:

- An identification mark code or number.
- The date and reference of maker's original test certificate.
- Date of commissioning.
- Dates of statutory tests, inspections and examinations.
- Details of defects and subsequent actions required.
- Copies of test certificates.

Thus, co-ordinated and well-documented procedures are obviously of paramount importance if items are not to escape inspection. The coding system as used in other contexts to identify a piece of equipment can be adopted for the inspection control purposes and should be painted on to the item itself, to aid identification. In addition, a colour coding for each period of statutory inspection helps to ensure that equipment meets the stipulated periods. Notwithstanding the voluntary or insurance company inspection procedures installed by a machine user or owner, the Health and Safety Executive may inspect equipment at any time and suspend its usage if defects are present and/or certificates are not up to date.

Maintenance and inspection records, etc., can be effectively monitored and updated using standard or customised computer packages mounted on personal, networked or mini computers.

9.3.4   Stock Control and Spare Parts Policy

Stock control can play an important role in securing the effectiveness of a maintenance system. Manufacturers, suppliers and transport systems are rarely able to deliver goods at the exact time required for the maintenance operation, and it is therefore necessary to carry sufficient stock to act as a buffer between supply and demand for a component. However, since the level of stock is only a buffer, it is important to keep levels to the minimum needed to service the maintenance requirements and so limit the locked-up capital, which otherwise could be more usefully employed elsewhere in the business.

The extent to which component types and stock levels are held will often depend upon the nature of the maintenance policy and the proximity of the manufacturers' distributors. For example, the fleet operator hiring out to the market would probably carry a sophisticated range of spares, whereas the company with only a few items would hold only those parts in frequent demand. Thus, the extent to which stock control is made effective is dependent upon:

(1) Defining a realistic stock objective in relation to the firm's activities.
(2) Using stock as a buffer only to aid production continuity.
(3) Setting economic levels of stock to service the needs of the enterprise.

It is not possible to offer firm advice in setting the correct stock control objectives, without having detailed information for a particular concern. However, the following techniques are available for dealing with the problems arising in items (2) and (3) above: (a) the ABC method of stock control; (b) inventory control.

The ABC Method

The ABC technique directs effort towards an ordering of stock priorities with the primary objective of avoiding stock-outs of critical items and keeping capital lock-up as low as possible. To operate the technique, stock is classified into three groupings:

'A' items are those which are most frequently demanded or expensive or which would have a significant impact if not provided.
'C' items are those not commonly requested, or which are inexpensive or would have the least effect if not provided.
'B' items are those which do not fall into categories 'A' or 'C'.

Clearly, for the stock control system to be effective, sophisticated information including forecasts of requirements, and lead times for ordering, will be required on 'A' items. 'B' items will require vigorous stock checking and frequent inspection, with more effort given to the high-priority items than to those of low priority. 'C' items need only routine checking, since the effect of a stock-out or miscalculation of the stock requirement will be relatively small.

Typically, stock items replenished over a common period could fall into the following categories:

'A' items: 10% in number, making up 70% of the value of stock.
'B' items: 20% in number, making up 20% of the value of stock.
'C' items: 70% in number, making up 10% of the value of stock.

ABC Stock Control Example

The spare parts department of a plant hire company currently replenishes all stock items every 3 months, as follows:

Class	Number of items	Value of stock (£)	ABC value of stock (£)
A	50	21 000	3 500
B	100	6 000	4 000
C	350	3 000	3 000
	500	30 000	10 500

The total cost of maintaining a uniform quantity of 3 months' minimum stock of all items is £30 000. However, with the items now clearly ranked into an order of stock priorities, decisions can be taken regarding more economic replenishing periods. For example, by keeping 2 weeks' stock of class 'A' items, 2 months' stock of class 'B' items and 3 months' stock of class 'C' items, the stock value reduces to £10 500. In this way capital is made available for other purposes.

It can be seen that the ABC method principally identifies the critical items, and so emphasis may be concentrated on checking class 'A' items to avoid stock-outs or supporting too much stock.

Inventory Control

While the ABC technique of stock control provides a simple checking method, it is usually also necessary to know when to order and the order quantity. In its simplest form, stock is ordered when the current level of stock, minus the immediate stock demand, equals the forecast demand before the next delivery arrives plus safety stock, thus:

$$\text{Current stock} - \text{immediate stock demand} = \underbrace{\text{forecast demand}}_{\text{order quantity}} + \text{safety stock}$$

The safety stock for a given lead time may be determined from past observations of differences between forecast demand and actual demand. The size of the forecast error, which may only be exceeded within a specified limit, defines the size of each safety stock to provide against a stock-out. Unfortunately, this method requires information on each stock item, which for a complex inventory would be too tedious to collect and so only Class 'A' items might be considered in this way.

Method of Determining the Safety Stock Level

(1) For each stock item calculate the forecast error by subtracting forecast demand from actual demand over a past time period. A positive difference represents a shortage, while a negative value indicates a surplus.

(2) Rank the range of forecast errors into 10–20 equal divisions — for example:

Range of errors	Forecast error frequency
− 24 to − 20 (items, quantity, etc.)	5 (occurrences)
− 19 to − 15	20
− 14 to − 10	40
− 9 to − 5	60
− 4 to 0	75
0 to + 4	80
+ 5 to + 9	65
+ 10 to + 14	40
+ 15 to + 19	15
+ 20 to + 24	5
	405

(3) A plot of the forecast errors against the frequency will tend to approximate to the normal distribution and simple statistics can be used to determine the probability of stock-outs and surpluses.

EXAMPLE 9.1

The forecast errors shown above were recorded for a particular stock item. Management wishes to know with 95% confidence the level of safety stock required to avoid a stock-out.

Solution

A plot of the forecast errors is shown in Figure 9.6. The sample is large and therefore approximates to a normal distribution, and the mean and standard deviation are calculated in the following table:

Range of errors	Mid-point of range, x	x − x̄	Frequency (f)	f × x	f × (x − x̄)²
− 24 to − 20	− 22	− 22.1	5	− 110	2 442
− 19 to − 15	− 17	− 17.1	20	− 340	5 848
− 14 to − 10	− 12	− 12.1	40	− 480	5 856
− 9 to − 5	− 7	− 7.1	60	− 420	3 024
− 4 to 0	− 2	− 2.1	75	− 150	331
0 to + 4	+ 2	+ 1.9	80	+ 160	289
+ 5 to + 9	+ 7	+ 6.9	65	+ 455	3 094
+ 10 to + 14	+ 12	+ 11.9	40	+ 480	5 664
+ 15 to + 19	+ 17	+ 16.9	15	+ 255	4 284
+ 20 to + 24	+ 22	+ 21.9	5	+ 110	2 352
			405	− 40	33 184

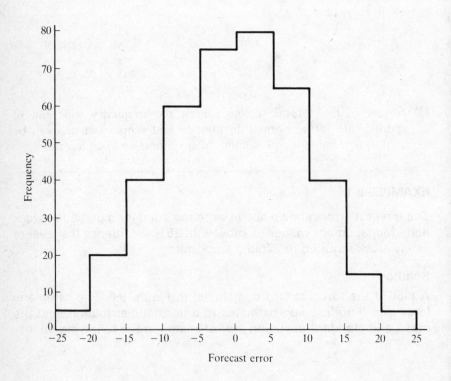

Figure 9.6   Distribution of forecast errors

$$\text{Mean stock error } (\mu) = \frac{\Sigma fx}{\Sigma f} = \frac{-40}{405} = -0.099$$

$$\text{Standard deviation } (\sigma) = \sqrt{\frac{\Sigma f(x - \bar{x})^2}{\Sigma f - 1}} = \sqrt{\frac{33\ 184}{404}} = 9.06$$

The data described by the normal distribution is governed by

$$Z = \frac{x \pm \mu}{\sigma} = \frac{\text{safety stock}}{\sigma}$$

From statistical tables, for a given error value x to lie within 95% of the area under the graph, $Z = 1.96$ — i.e. within 1.96 standard deviations of the mean. Therefore, safety stock $= 1.96 \times 9.06 = 17.76$ (say 18). If stock were ordered in 4 week cycles, the probability is that a stock-out will occur every 80 weeks — i.e. 5% chance.

Economic Order Quantity (How Much to Order?)

While the order quantity formula already given is appropriate for simple situations, for more complex stockholdings the costs of procurement and storage can be high, in such circumstances a more general formula is necessary relating order quantity, order period and costs. Two costs must be considered:

(1) *Cost of procurement* Administration costs are incurred whenever an order is placed. These will include (a) purchase enquiries, requisitions and ordering; (b) acceptance, inspection and legal requirements; (c) preparation of drawings and design details, etc.
(2) *Cost of storage* This includes (a) interest to be paid on working capital; (b) insurance; (c) storage and handling; (d) maintenance records; (e) wastage, theft, deterioration, etc.

The most economical ordering quantity involves balancing these costs against the rate of usage.

EXAMPLE 9.2

Stocks of a component are allowed to run down to a level of three units before being replenished. The components are used steadily at 50 items per week. The component costs £15 per unit and the cost of storage and deterioration per week is 10% of the cost price. Each time a component is ordered there is a cost of processing this order of £1.

(1) How many items should be ordered each time?
(2) What is the cost of ordering and storing each item?

Solution

Let B = minimum stock level — i.e. safety stock. Let Q = number of components delivered with each order; D = rate of usage in units per week; S = cost of processing an order; h = cost of storing an item per week as a percentage of cost price; P = cost of an item, t = time in weeks between orders.

The cycle of usage and replenishment is shown in Figure 9.7.

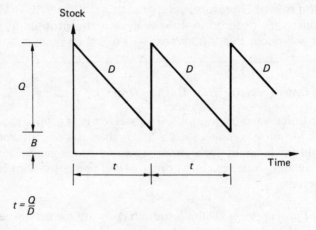

Figure 9.7

Average number of items stored in time t is $\frac{1}{2}Q + B$. Therefore, storing cost per cycle of length $t = \frac{1}{2}QthP + BthP$.

Total cost per cycle of length $t = \frac{1}{2}QthP + BthP + S$

Total cost per week $= \frac{1}{2}QhP + BhP = \dfrac{S}{t}$

But $t = Q/D$.

$\therefore$ Total cost (TC) per week $= \frac{1}{2}QhP + BhP + \dfrac{SD}{Q}$ \hfill (1)

To obtain optimum order size, differentiate with respect to Q:

$$\frac{dTC}{dQ} = \tfrac{1}{2}hP - \frac{SD}{Q^2}$$

$$= 0 \text{ for a maximum}$$

Therefore

$$Q^2 = \frac{2SD}{hP}$$

(1)

$$Q = \sqrt{\frac{2SD}{hP}}$$

$$= \sqrt{\left[\frac{2 \times 1 \times 50}{(1/10) \times 15}\right]}$$

$$= 8.16 \text{ components} \tag{2}$$

(2) Substituting Q in Equation (1)

$$TC = hPB + \frac{Q^2 hP + 2SD}{2Q}$$

$$= hPB + \frac{2SD + 2SD}{2 \times \sqrt{\frac{2SD}{hP}}}$$

$$= hPB + \sqrt{2SDhP} \tag{3}$$

$$= (1/10) \times 15 \times 3 + \sqrt{[2 \times 1 \times 50 \times (1/10) \times 15]}$$

$$= £16.75 \text{ per week}$$

EXAMPLE 9.3: STOCK CONTROL AND SHORTAGES

A rental firm calls for the steady supply of 50 components each week; the price of the component is £30. The supplier usually keeps sufficient stocks to meet demand, and the cost of holding a component per week is 10% of cost price. The cost to the supplier each time a new order is processed is £10. However, sometimes the delivery date cannot be met, so to make up the backlog there are special deliveries to the customer as soon as the supplier is able to continue with the order. The extra cost incurred by the supplier in this situation is £10 per component.

(1) Calculate the economic order quantity for the supplier.
(2) Calculate the total cost per week to the supplier of stockholding and processing orders.
(3) Calculate the level to which stock on site is topped up.

Solution

Z = cost of shortage per component, Q = economic order quantity; A = top-up quantity; D = rate of usage per week; h = storage cost as a percentage of the cost price of component; P = cost of components; S = cost of processing an order; t = time in weeks between supplies.

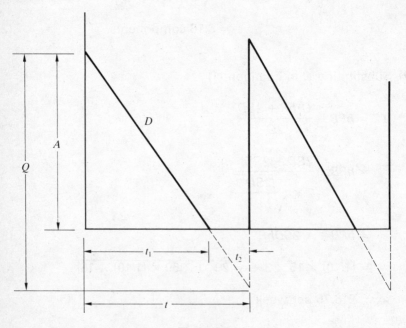

Figure 9.8

From Figure 9.8,

(a) Storing cost = $\frac{1}{2}Dt_1t_1hP$ per cycle and $A = Dt_1$

Storing cost per week = $\dfrac{Dt^2_1hP}{2t}$

$$= \frac{At_1 hP}{2(t_1 + t_2)} \tag{1}$$

Now, from similar triangles,

$$\frac{Q}{t_1 + t_2} = \frac{A}{t_1}$$

Therefore,

$$t_1 = \frac{At_2}{Q - A}$$

Substituting t_1 in Equation (1),

$$\text{Storing cost per week} = \frac{A^2 hP}{2Q} \tag{2}$$

(b) Cost of shortage per cycle $= \frac{1}{2} Dt_2^2 Z$

$$\text{Shortage cost per week} = \frac{DT_2^2 Z}{2t}$$

$$= \frac{Dt_2^2 Z}{2(t_1 + t_2)} \tag{3}$$

Substituting t_1 in Equation (3),

$$\text{Shortage cost per week} = \frac{(Q - A)^2 Z}{2Q} \tag{4}$$

(c) Total cost per week $= \dfrac{A^2 hP}{Q} + \dfrac{(Q - A)^2 Z}{2Q} + \dfrac{SD}{Q}$

To obtain optimum order size, differentiate with respect to Q and A and maximise.

$$Q = \sqrt{\frac{2SD}{hP}} \cdot \sqrt{\frac{Z + hP}{Z}} \text{ units}$$

$$A = \sqrt{\frac{2SD}{hP}} \cdot \sqrt{\frac{Z}{Z + hP}} \text{ units}$$

$$TC = \sqrt{2SDhP \cdot} \sqrt{\frac{Z}{z + hP}} \text{ per week}$$

(1)

$$Q = \sqrt{\left[\frac{2 \times 10 \times 50}{(1/10) \times 30}\right]} \times \sqrt{\left[\frac{10 + (1/10) \times 30}{10}\right]}$$

$$= 20.8 \text{ components}$$

(2)

$$TC = \sqrt{[2 \times 10 \times 50 \times (1/10) \times 30]} \times \sqrt{\frac{10}{10 + (1/10) \times 30}}$$

$$= £48.04 \text{ per week}$$

(3)

$$A = \sqrt{\left[\frac{2 \times 10 \times 50}{(1/10) \times 30}\right]} \times \sqrt{\left[\frac{10}{(10) + (1/10) \times 30}\right]}$$

$$= 16.01 \text{ components}$$

EXAMPLE 9.4: STOCK CONTROL AND DISCOUNTS

Components are required at the rate of 2 000 per month. The cost of ordering is £20 and the cost of storing the material is 50% of its purchase cost. The cost per item depends on the total quantity ordered as follows: (1) less than 500 items @ £1.21 per item; (2) 500–999 items @ £1.00 per item; (3) 1 000 or more @ £0.81 per item.
   Calculate the optimum order quantity and the optimum total cost per month of purchasing, storing and ordering the material.

Solution

$$Q = \sqrt{\frac{2SD}{hP}}$$

$$= \sqrt{\frac{2 \times 20 \times 2000}{\frac{1}{2}P}}$$

$$= \frac{400}{\sqrt{P}}$$

Without discounts, only the costs of storage have been involved in the calculation and not the cost of the material itself. This may result in a false situation, as follows:

(1)
$$Q = \frac{400}{\sqrt{1.21}}$$

$$= \frac{400}{1.1}$$

$$= 363 < 500 \text{ items (i.e. within range)}$$

(2)
$$Q = \frac{400}{\sqrt{1}}$$

$$= 400 \text{ (i.e. outside range 500–999 items).}$$

(3)
$$Q = \frac{400}{\sqrt{0.81}}$$

$$= \frac{400}{0.9}$$

$$= 444 \text{ (i.e. outside range 1000 items or more)}$$

On first inspection the optimum order quantity would be 363 units, but the calculation so far does not take into account the cost of the material itself, which in this instance varies according to the quantity ordered. Therefore, looking at total costs:

$$TC = \text{cost of material} + \text{storage cost} + \text{order cost}$$

Therefore,

$$TC \text{ per month} = DP + \tfrac{1}{2}QhP + \frac{SD}{Q}$$

Overall discount situation:

(1)
$$TC = 2\,000 \times 1.21 + (\tfrac{1}{2} \times 363 \times 0.5 \times 1.21) + \frac{20 \times 2\,000}{363}$$

$$= £2\,640 \text{ per month}$$

(2)

(a)

$$TC = 2\,000 \times 1.0 + (\tfrac{1}{2} \times 500 \times 0.5 \times 1.0) + \frac{20 \times 2\,000}{500}$$

$$= £2\,205 \text{ per month}$$

(b)

$$TC = 2\,000 \times 1.0 + (\tfrac{1}{2} \times 999 \times 0.5 \times 1.0) + \frac{20 \times 2\,000}{999}$$

$$= £2\,289 \text{ per month}$$

(3)

(a)

$$TC = 2\,000 \times 0.81 + (\tfrac{1}{2} \times 1\,000 \times 0.5 \times 0.81) + \frac{20 \times 2\,000}{1\,000}$$

$$= £1\,862 \text{ per month}$$

(b)

$$TC = 2\,000 \times 0.81 + (\tfrac{1}{2} \times 2\,000 \times 0.5 \times 0.81) + \frac{20 \times 2\,000}{2\,000}$$

$$= £2\,045 \text{ per month}$$

The optimum order quantity, therefore, is 1 000 items per month and the total monthly purchase, storage and ordering cost is £1 862.

EXAMPLE 9.5: STOCK CONTROL AND CONTINUOUS USAGE

A manufacturer is required to supply 1 000 parts each week to replenish stocks. The store has very little storage space and thus requires the units to be delivered at the rate at which they can be used. The manufacturer has the capacity to produce 2 500 units per week. The cost of storing a unit per week is 1p and the cost of setting up the equipment for a production run is £50.

(1) What is the optimum number of units to produce in a production run?

(2) What is the total cost of producing and storing the plant department's requirements?

(3) How frequently should production runs be made?

Solution

(1)

Q = number of units made per production run, D = number of units required by contractor each week, k = number of units produced per week, H = cost of storing one item per week, S = cost of setting up a production run, t = time interval in weeks between production runs. From Figure 9.9,

Length of the production run $t_1 = \dfrac{Q}{k}$

Length of production and usage cycle $t = \dfrac{Q}{D}$, $AC = Q - Dt_1$

Storage cost per cycle $= \frac{1}{2}(Q - Dt_1) \times H \times t = \frac{1}{2}Qt\left(1 - \dfrac{D}{k}\right)H$

Total cost per cycle $= \frac{1}{2}QtH\left(1 - \dfrac{D}{k}\right) + S$

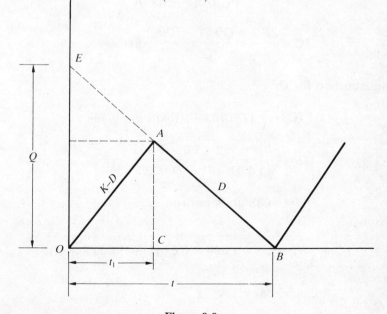

Figure 9.9

Therefore

$$\text{Total cost (TC) per week} = \frac{S}{t} + \tfrac{1}{2}QH\left(1 - \frac{D}{k}\right)$$

$$TC = \frac{SD}{Q} + \tfrac{1}{2}QH\left(1 - \frac{D}{k}\right)$$

To obtain optimum run size, differentiate with respect to Q:

$$\frac{dTC}{dQ} = -\frac{SD}{Q^2} + \tfrac{1}{2}H\left(1 - \frac{D}{k}\right)$$

$$= 0 \text{ or a maximum}$$

$$Q = \sqrt{\frac{2SD}{H[1 - (D/k)]}}\,\text{units}$$

$$t = \sqrt{\frac{2S}{DH[1 - (D/k)]}}\,\text{weeks}$$

$$TC = \frac{2SD + Q^2H[1 - (D/K)]}{2Q}\,\text{per week}$$

or, substituting for Q,

$$TC = \sqrt{[2SDH[1 - (D/k)]]}\,\text{per week} \qquad (2)$$

$$Q = \sqrt{\left[\frac{2 \times 50 \times 1\,000}{0.01 \times [1 - (1\,000/2\,500)]}\right]}$$

$$= 4\,083 \text{ units per run}$$

(2)

$$TC = \sqrt{(2 \times 50 \times 1\,000 \times 0.01\,[1 - (1\,000/2\,500)])}$$

$$= £24.50 \text{ per week}$$

(3)

$$t = \sqrt{\frac{2 \times 50}{1\,000 \times 0.01\,[1 - (1\,000/2\,500)]}}$$

$$= 4.08 \text{ weeks}$$

These examples clearly involve a refined level of stock control, necessitating detailed information gathering. For many equipment stores departments, the types of stocks and spare parts are too few, and usage too low, to need such sophistication. In that circumstance, adoption of the ABC technique coupled with rigorous checking of both stock and order processing should suffice.

BIBLIOGRAPHY

Bailey, P. J. H. (1987). *Purchasing and Supply Management*, Chapman and Hall

Bailey, P. J. H. and Tavernier, G. (1984). *Stock Control Systems and Records*, Gower

Battersby, A. (1962). *A Guide to Stock Control*, Pitman

Clifton, R. H. (1974). *Principles of Planned Maintenance*, Edward Arnold

Construction Equipment Maintenance, Cahners, Chicago (monthly)

Corder, A. S. (1976). *Maintenance Management Techniques*, McGraw-Hill

Cordero, S. T. (1989). *Maintenance Management Handbook*, Fairmont Press

Higgins, L. R. (1979). *Handbook of Equipment Maintenance*, McGraw-Hill

Priel, V. F. (1975). *Systematic Maintenance Organisation*, McDonald and Evans

Chapter 10
Health and Safety at Work: Regulations and Requirements

10.1 BASIC PROVISIONS OF THE HEALTH AND SAFETY AT WORK ACT

The Health and Safety at Work Act, introduced in 1974, provided a legislative framework designed to encourage high standards of health and safety at work. The Act is described as an enabling measure superimposed over existing health and safety legislation contained in some 31 Acts. The existing legislation, mainly under the Factories, Offices, Shops and Railway Premises and Mines and Quarries Acts, remains in force, but the Health and Safety at Work Act provides one comprehensive and integrated system of law relating to health and safety at work. The Act consists of four parts: *Part I*, relating to health, safety and welfare in relation to work; *Part II*, relating to an Employment Medical Advisory Service; *Part III*, relating to amendments in the law relating to building regulations; and *Part IV*, containing a number of miscellaneous and general provisions.

The Health and Safety at Work Act was intended to provide for the gradual replacement of the then existing health and safety requirements by revised and updated provisions. These new provisions would be in the form of regulations and codes of practice prepared in consultation with industry.

The regulations are made by the appropriate Minister and, where appropriate, are supplemented by codes of practice which, although not statutory, can be used as evidence that the statutory requirements have been contravened.

The Act provided for the establishment of a Health and Safety Commission and a Health and Safety Executive.

10.1.1   Health and Safety Commission

The Health and Safety Commission has a full-time chairman and nine lay members. It is responsible for developing the policies in Health and Safety. The Commission consists of representatives of industry, unions and local authorities, and is ultimately responsible to the Minister for Employment.

10.1.2   Health and Safety Executive

The Health and Safety Executive is a separate statutory body appointed by the Commission to work within the direction and guidance of the Commission. The main duty of the Executive is to enforce the legal requirements and to provide an advisory service. The Executive is responsible for the Health and Safety Inspectorates (i.e. Factories, Mines, Quarries, Construction, etc.).

10.1.3   Health and Safety Inspectorate

The Act's major legal requirements are enforced through the work of the inspectors. If an inspector discovers a contravention, he is empowered to issue a prohibition notice or an improvement notice, to prosecute or to seize, render harmless and destroy a dangerous article.

A prohibition notice is issued if there is a serious risk to health or a risk of serious injury, and the activity giving rise to that risk must be stopped until remedial action can be taken. The prohibition notice is served on the person undertaking the activity or on the person in control.

An improvement notice is issued so that a fault or contravention of a statutory requirement can be remedied within a specified time. The notice would be served on the person who is not satisfying the legal requirement or on any person on whom the responsibility has been placed, and this could be an employer, an employed person or even a supplier of equipment.

There is the right of appeal to an independent tribunal against both a prohibition notice and an improvement notice. Prosecution can take place instead of or in addition to serving a notice. Failure to comply with a notice is also an offence against the law. The inspector can also seize, render harmless and destroy any substance or article that is the cause of imminent danger or serious personal injury.

Inspectors may also take written statements of fact from witnesses and ultimately conduct proceedings in court even though they are not solicitors or counsel.

10.1.4   Responsibilities and Duties

The Health and Safety at Work Act also defines the responsibilities and duties of the various parties.

Employers

The duties of the employer include:

(1) Ensuring, as far as is reasonably practicable, the health, safety and welfare of all employees. This applies particularly to the provision of equipment and systems of works that are safe and without risks to health.
(2) Ensuring that all machinery, equipment and appliances used are safe.
(3) Ensuring that the handling, transport and storage of materials are safe.
(4) Providing information, instruction, training and supervision to ensure the health and safety of employees at work.
(5) Maintaining safe means of access to and egress from places of work.
(6) Provision and maintenance of a safe working environment with adequate facilities for welfare at work.
(7) Preparation and revision of a written statement explaining the employer's general policy with respect to the health and safety of the employees and the organisation and arrangements made to carry out that policy.

Employers also have a duty with regard to the health and safety of persons who are not employees, such as self-employed or contractors' employees who may be working close to their own employees. This duty extends to members of the public and, employers must ensure that the public are not exposed to health and safety risks.

Employees

The Act places on the employees the duty to take reasonable care to avoid injury to themselves or to others as a result of their activities at work. Employees must not interfere with or misuse anything provided to protect their health and safety, and must also co-operate with their employer in respect to health and safety matters.

Manufacturers and Suppliers

The Act places on manufacturers and suppliers the duty to ensure that any article or substance supplied for use at work is safe when properly used in a reasonable manner.

Thus, an equipment hire company has the responsibility placed upon it to ensure that any equipment supplied is safe when properly used. The supplier must test any article (i.e. item of equipment) for safety in use or arrange for a competent authority to test the article. The supplier must also supply information about the use for which an article was designed and include conditions of use regarding its safety.

The general obligations imposed by the Health and Safety at Work Act 1974 are supplemented by the detailed requirements of regulations. The following regulations are specifically of interest to the construction industry: Construction (General Practices) Regulations 1961; Construction (Lifting Operations) Regulations 1961; Construction (Working Places) Regulations 1966. In addition, there are other regulations dealing with specific hazards, of which the following are examples relevant to the construction industry: asbestos; work in compressed air; abrasive wheels; and woodworking machinery.

10.1.5 Safety Statistics

The Health and Safety Executive's report records the safety statistics. Table 10.1 presents the incidence of fatal and major injuries experienced in the UK construction industry in the years 1981–1985.

Table 10.1 Number of injuries per 100 000 employees in construction (from H. M. Chief Inspector of Factories, 1986/87)

	1981	1982	1983	1984	1985
Fatal injury rate	9.7	9.9	11.4	9.8	10.8
Combined fatal and major injury rate	164	204	221	235	237

Table 10.2 Fatal accidents in construction by cause

Cause	1981 No.	1981 %	1982 No.	1982 %	1983 No.	1983 %	1984 No.	1984 %	1985 No.	1985 %
Falls	73	57	77	57	95	63	75	57	84	61
Transport	27	21	27	20	18	12	27	21	23	16.5
Electrocution	7	5.5	6	4.5	11	7.5	6	4.5	4	3
Fire and explosion	0	0	2	1.5	3	2	2	1.5	3	2
Collapse	7	5.5	10	7.5	10	6.5	13	10	13	9.5
Insecure loads	9	7	8	6	6	4	5	4	9	6.5
Drowning	3	2.5	5	3.5	8	5	2	1.5	2	1.5
Unclassified	2	1.5	0	0	0	0	1	0.5	0	0
	128		135		151		131		138	

The total numbers and cause of the fatal accidents are presented in Table 10.2. This shows that falls remain the most common cause.

10.2   SAFETY POLICY AND ORGANISATION

Because construction sites are remote from head office and because construction sites have to contend with a widely varying number of factors such as climate, geology, type of work, etc., the role of the company's safety policy and how it is implemented is particularly important. The company's safety policy needs to be supported by realistic organisational arrangements that will make the implementation of the policy effective. There needs to be:

(1) A clear and logical delegation of duties through line management.
(2) Clear identification of key personnel to carry out the detailed arrangements, such as the repair, maintenance and inspection of equipment, including the keeping of the required records.
(3) Clear job descriptions and organisation charts which show the relationship between line managers and functional managers (such as safety officers), demonstrating the responsibility each manager has for safety.
(4) A well-defined mechanism to ensure that the safety officer monitors the health and safety aspects of the company operations and reports back to senior managers in a manner in which a clear statement of the health and safety at each site is given.

The company's policy and organisational arrangements must be such that safe systems of work are developed and used. In most construction companies the method or system of work is devised by the site management team. These are generally called method statements. The personnel responsible for developing these method statements must be aware of the safety requirements.

The company also has a duty to train personnel. This means not only operatives who have to be trained in the correct use of equipment, but also the supervisors and site managers who are responsible for devising safe methods of work.

Under the Control of Substances Hazardous to Health Regulations (COSHH) 1988, employers are required to assess risks created by work in which a hazardous substance is used. The employer is required to

• Assess the risks to people's health, from the way in which a substance is used.
• Determine whether exposure to the substance can be prevented.
• Determine how to control the exposure and reduce the risk.

- Establish effect controls.
- Train and inform the workforce.
- Provide health surveillance.

10.2.1   Safety Representatives

Where two or more members of a recognised trade union request an employer to form a joint Health and Safety Committee, the employer is required to do so by the regulations on Safety Representatives and Safety Committees, SI 1977 No. 500.

The elected Safety Representative is entitled to:

(1) Investigate accidents, hazards and complaints and make representations to the employer.
(2) Conduct health and safety inspections.
(3) Receive information from Health and Safety Inspectors.
(4) Attend meetings of the Health and Safety Committee.
(5) Receive information from the employer in respect of health and safety issues and inspect statutory documents.

The Safety Representative is also permitted time off with pay, to perform the above tasks and also to receive adequate training.

10.3   NOTIFICATION OF ACCIDENTS AND DANGEROUS OCCURRENCES

The current regulations for the notification of accidents and dangerous occurrences have been effective since 1 April 1986. The previous requirements were revoked.

Accidents and dangerous occurrences can be classified as: minor injuries; major injuries; fatal accidents; road vehicle accidents; dangerous occurrences. The notification requirements of each of these classifications is described below.

10.3.1   Minor Injuries

Minor injuries are defined as injuries which keep the employee from work for more than 3 consecutive days. The employer is required to notify the Department of Social Security on Form B176, when it is received, after a claim by the employee for industrial injury benefit. The DSS will then send particulars of the accident to the Health and Safety

Executive. If there is no claim for benefit, the employer need not report the accident but must keep a record of the accident.

10.3.2   Major Injuries

A major injury is defined as a fracture of skull, shin, pelvis or any bone, amputation, loss of sight or any injury which results in hospitalisation for more than 24 hours.

The employer is required to notify the enforcing authority by the quickest means possible, usually by telephone. The employer must send a report on Form 2508 to the enforcing authority within 7 days, and the same notification procedure is required for persons other than the company's employees, such as members of the public, involved in a major accident.

10.3.3   Fatal Accident

The notification procedures for an accident resulting in the death of a person are the same as for major injuries. Form 2508 should be used for reporting to the authorities.

10.3.4   Road Vehicle Accidents

Accidents involving road vehicles adjacent to the public highway are notifiable only if the person was engaged at the time in work on or alongside a road.

10.3.5   Dangerous Occurrences

Specified dangerous occurrences which have the potential for causing serious accidents even though physical injuries may not have occurred are notifiable in the same way as major injuries and fatalities. Examples of dangerous occurrences are scaffold collapse, overturning of a crane, explosions, or collapses of structures and buildings under construction.

10.3.6   Records

Employers are required to keep written records of all notifiable accidents and dangerous occurrences. The details required in these records, which should be kept for 3 years, are: the date of accident or dangerous occurrences; in the case of accidents, the name, age, sex, occupation of the victim and nature of injury; the place where the accident or dangerous occurrence took place; a brief description of the circumstances.

10.4   SAFETY CHECK LIST

The Health and Safety Executive have produced a basic checklist for on site. This list of safety points was produced by the Inspectorate in the light of their experience. The list was intended as an aid to safety supervisors, safety representatives and site managers. This check list is reproduced here with the kind permission of the Controller of Her Majesty's Stationery Office.

10.4.1   Construction Site Basic Safety Check List

Is It Safe?

One sure way to reduce accidents is to pinpoint unsafe situations or practices, so that steps can be taken to correct them before anyone gets hurt. An attempt has been made to identify some of the most elementary hazards found on sites, which frequently lead to accidents. This check list, as its title says, is basic and it should therefore be modified to suit local needs.

Safe Access

More than 50% of the accidents that keep men away from work involve falls or collisions, of men, materials and vehicles. It is therefore vital that access from place to place be made safe.

Check Points

(1) Is safe access provided for all on site to reach their places of work — i.e. good roads, gangways, passageways, passenger hoists, staircases, ladders and scaffolds?
(2) Are all walkways level and free from obstruction?
(3) Is edge protection provided or other preventive measures taken where men are likely to fall from an open side?
(4) Are holes or openings covered over with securely fixed covers or, alternatively, fenced off?
(5) Is adequate artificial lighting available for when work has to continue after dark?
(6) Is the site tidy and are materials stored in safe positions?

(7) Are there proper arrangements for the gathering and disposal of scrap?

(8) Are nails in timber hammered down or removed?

Ladders

More accidents arise each year from the use and/or misuse of ladders than of any other single piece of equipment.

Checkpoints

(1) Is every ladder in good condition and free from obvious defects?

(2) Are all ladders secured near the top, including those used for short periods?

(3) If the ladder cannot be secured near the top, is it secured near the bottom, weighted or footed?

(4) Does the ladder rise at least 1.07 m (3 ft 6 in) above the place of landing? If not, is there adequate handhold at the place of landing?

(5) Are ladders properly positioned for access?

Tubular Scaffolds

The law requires that scaffolding work be done with competent and experienced supervision. All scaffolders except trainees should be experienced and competent in their work. The loading for which a scaffold has been provided should be known. The loading should be evenly distributed. The scaffold should not be overloaded. Scaffolds are required to be inspected at least once a week and after bad weather.

Checkpoints

(1) Has proper access been provided to the scaffold platform?

(2) Are all uprights provided with base plates or prevented in some other way from slipping or sinking?

(3) Have any uprights, ledgers, braces or struts been removed?

(4) Is the scaffold secured to the building in enough places to prevent collapse? Have any ties been removed since the scaffold was erected?

(5) Are there sufficient boards at all working platforms in use?

(6) Are all boards free from obvious defects, and are they arranged to avoid traps?

(7) Are there warning notices to prohibit the use of any scaffold that is incomplete — e.g. not fully boarded?

(8) At every side where a person can fall more than 1.98 m (6 ft 6 in) are the platforms, gangways and runs provided with guardrails and toeboards?

(9) Who is responsible for the inspections, and are they carried out and recorded?

Excavations

The digging of an excavation on a site may be a man's first and last job. If the sides collapse, there may be no escape. There is almost no ground which will not collapse under certain conditions. If there is any doubt whatsoever about the safety of the sides, they should be timbered or battered. Timbering materials should always be provided on site.

If the excavation is more than 1.21 m (4 ft) deep then the following checkpoints apply:

(1) Is the timber provided strong enough to support the sides of the excavation?
(2) Is the method of putting in timbering a safe one?
(3) Is the angle of batter appropriate?
(4) Is the excavation inspected daily, and the timbering weekly?
(5) Is there safe access to the excavation?
(6) Is there a barrier to prevent persons falling in?
(7) Is the stability of the excavation being affected by vehicles which come too near?
(8) If vehicles tip into an excavation, are properly secured stop blocks provided?

Roof Work

Checkpoints

(1) Is work being done on a sloping roof of more than 30° or less than 30° but which is slippery? If so, are there crawling ladders or crawling boards? Are these being used? If crawling boards or ladders are not being provided, does the roof structure itself provide a safe handhold and foothold?
(2) For sloping roofs or work near the edge of flat roofs, is there sufficient edge protection to prevent falls of materials and persons?
(3) Are any men working near or on fragile materials such as asbestos cement sheets or glass? If so, are crawling boards provided and used and warning notices posted?
(4) Have all rooflights been properly covered or provided with barriers?
(5) During sheeting operations are precautions taken to prevent men falling from the edge of the sheet?
(6) Where other men are working under roofwork are precautions taken to prevent debris falling onto them?

Transport

Checkpoints

(1) Are vehicles kept in good repair?
(2) Is the steering satisfactory and are the handbrake and footbreak working properly?
(3) Are vehicles driven in a safe way?
(4) Do the vehicles have any parts likely to cause injury in the normal circumstances of their use?
(5) Are vehicles badly loaded so that there is danger from falling loads?
(6) Do persons ride in dangerous positions?
(7) Are there any tipping lorries? If so, are the dangers of reaching under the raised body understood?

Machinery

Some dangerous parts, such as gears, chain drives and rotating shafts with projecting keys and set-screws, are easy to recognise. Others are not so obvious, such as projecting engine shafts.

Checkpoints

(1) Are there any dangerous parts?
(2) Are they guarded?
(3) Is the guard secured and in good repair?

Hoists

A platform hoist must be protected by a substantial enclosure and the enclosure must be fitted with gates where access is needed. A platform hoist can be dangerous if the gates are not kept shut. Hoists are required to be inspected weekly, and thoroughly examined every 6 months by a competent person.

Checkpoints

(1) Who is responsible for these inspections, and are they carried out and recorded?
(2) Is there an enclosure, where necessary, to prevent people being stuck by any moving part of the hoist or materials on it?
(3) Are gates provided at all landings?
(4) Are the gates kept shut except when the platform is at the landing?
(5) Is the control rope so arranged that the hoist can be operated from one position only?
(6) Is the safe working load clearly marked on the hoist?
(7) Is there a proper signalling system?

(8) If the hoist is for materials only, is there a notice on the platform or cage prohibiting persons from riding on it? Is this notice obeyed?

Cranes and Lifting Appliances

The collapse and overturning of a crane can injure other people as well as the crane driver, especially on a crowded site or where the crane is working near a public thoroughfare. Cranes are required to be inspected weekly, and thoroughly examined every 14 months by a competent person.

Checkpoints

(1) Who is responsible for these inspections, and are they carried out and recorded?
(2) Has the test certificate been sent?
(3) Is the driver trained and competent and over 18 years old?
(4) Are the controls (levers, handles, switches, etc.) clearly marked?
(5) Has the weight of the load been ascertained?
(6) Is the crane of more than 1 ton capacity? If so, and it is a jib crane, is it fitted with an automatic safe load indicator in efficient working order, which is being inspected weekly?
(7) Is the crane sited on a hard level base?
(8) Has the slinger been trained to give signals and to attach loads correctly?
(9) If the crane can vary its operating radius, are the safe working loads and corresponding radii plainly marked?
(10) Is the crane regularly maintained?

Electricity

The main causes of electrical accidents on sites are from electrical equipment or lighting, from overhead electric power lines and from underground cables. Treat all electric equipment with respect.

Checkpoints

(1) Signs of damage to apparatus — especially portable equipment.
(2) Signs of damage to outer covering of wires and cables.
(3) Are all connections to power points by proper plugs, etc., and not bare ends of cables?
(4) Are there signs of interference with equipment, damaged or otherwise?
(5) Are there any overhead electric lines? If so, is there anything (a crane, forklift truck, tipper lorry, excavator or scaffolding) which might touch these lines? If so, has the electricity supply to these lines been cut off?
(6) If not, what other precautions have been taken?

(7) If there is an electric underground cable in the vicinity of work being
     carried out, is the line of the cable known, has it been located and
     marked and have precautions been taken to prevent contact?

Manual Handling

Ignorance of the best way of lifting a load is a common cause of accidents.
Lift with the legs bent, and the back straight. Get a good grip; use gloves if
there are any sharp edges; do not lift if the weight is too great.

Trespassers

Sites should always be made as secure as possible against trespass by
children especially at times when no one is on the site.

Checkpoints

(1) Are all ladders removed at the end of each working period or made
     incapable of use by boarding the rungs?
(2) Is all equipment immobilised at the end of each working period?
(3) Are bricks and materials safely stacked?
(4) If there is perimeter fencing, is it undamaged and are the gates
     secured?

Health Risks

Risks to health, which may not show themselves immediately, can arise
from some materials. Examples are asbestos, spraying of certain types of
plastic paints, burning or cutting lead-painted materials, cleaning of
sandstone buildings and the use of sand for cleaning buildings. Confined
spaces such as manholes and sewers can be dangerous because of lack of
oxygen or the presence of fumes and dust.

Checkpoints

(1) Have harmful materials been identified?
(2) What precautions are needed?
(3) Is the necessary safety equipment provided and properly used?
(4) Are other workers, who are not protected, kept out of danger areas?
(5) In confined spaces has the atmosphere been tested and an air supply
     provided, if necessary?

Protective Clothing

The head, eyes, hands and feet are all very vulnerable to injury. Equipment to prevent accidents can be made available and eye protection for
certain dust- or particle-creating processes is required by law.

Checkpoints

(1) Is protective equipment provided at least to the minimum standard required by law?
(2) Do persons employed wear their protective equipment?

Fire

Fires on construction sites can occur from the ignition of highly flammable liquids used on site — e.g. adhesives, and floor and wall coatings. The misuse of compressed gases has also been the cause of fire and explosions on sites, and rules for storage and use of cylinders should be strictly observed. Other causes of fire are accidental ignition of waste material, wood shavings and cellular plastic materials used as insulation or cavity fill material.

Checkpoints

General:
(1) Are there any fire extinguishers?
(2) Is there a secondary way out if fire blocks the normal route?
Highly flammable liquids:
(3) Is there a properly constructed and sited store area?
(4) Is the amount of flammable liquid present at the workplace kept to the minimum for the day's work?
(5) Is smoking prohibited, and are other ignition sources excluded from the areas where the liquids are present?
(6) Are properly constructed safety containers in use?
Compressed gases such as LPG and acetylene:
(7) Are the cylinders properly stored?
(8) Is the cylinder valve fully closed when the cylinder is not in use?
(9) Are cylinders sited outside huts, with a piped supply into the hut?
Other combustible material:
(10) Are proper waste receptacles provided?
(11) Is provision made for regular removal of waste material?

Explosives

A competent person with knowledge of the dangers should have charge of the storage, handling and use of explosives.

Cartridge-operated Tools

The maker's instructions should always be carefully followed.

Checkpoints

(1) Has the man using the gun been properly trained and told about the dangers?
(2) Is he wearing goggles?
(3) Does he know how to deal with misfires?
(4) Is the gun being cleaned regularly?
(5) Is the gun kept in a secure place when not in use?
(6) Are cartridges kept in a secure place?

Noise

Noise can affect men's hearing if they are exposed to it for long periods.

Checkpoints

(1) Is there any uncomfortably noisy plant or machinery?
(2) Are breakers fitted with muffs and other plant or machinery fitted with silencers?
(3) In very noisy surroundings are ear defenders supplied to the men?
(4) Do people have to work in places where they must shout to be clearly audible to someone 1–2 m away?

Falsework/Formwork

Temporary works can be a source of danger if not properly constructed.

Checkpoints:

(1) Has anyone checked the design and supports for the shuttering and formwork?
(2) Are the props plumb and properly set out?
(3) Are the bases and ground conditions adequate for the loads?
(4) Are the proper pins used in the props?
(5) Are the timbers in good condition?

Welfare

Checkpoints

(1) Are the lavatories and washbasins kept clean?
(2) Is the mess hut kept clean and free from storage?
(3) Is there a hut where wet clothes can be dried?
(4) Is there a supply of drinking water?
(5) Is there a first aid box?

10.5   GUIDE TO STATUTORY TESTS, EXAMINATION AND INSPECTIONS

In fulfilling the requirements of the Health and Safety at Work Act and the derived regulations there are a number of statutory tests, examinations and inspections that are required. These are summarised in Figure 10.1, which was prepared by Geo. Wimpey & Co. Ltd. and originally published in *Building Technology and Management* in March 1974, but the requirements have been checked for current accuracy. The statutory tests, examinations and inspections are particularly relevant to users of equipment on site.

This guide is reproduced by the kind permission of the Editor of *Building Technology and Management* and Geo. Wimpey & Co. Ltd.

10.6   SOME LEGISLATION ENCOMPASSED IN HEALTH AND SAFETY RELEVANT TO CONSTRUCTION

Health and Safety at Work Act, 1974

Factories Act, 1961

Building (Safety, Health and Welfare) Regulations, 1948 (SI 1948 No. 1145) (Regulations 1–4 and 99)

Construction (General Provisions) Regulations 1961 (SI 1961/1580)

Construction (Lifting Operations) Regulations 1961 (SI 1961/1581)

Construction (Working Places) Regulations 1966 (SI 1966/94)

Construction (Health and Welfare) Regulations 1966 (SI 1966/95), as amended by SI 1981/917

Engineering Construction (Extension of Definition) Regulations 1960 (SI 1960/421)

Engineering Construction (Extension of Definition) Regulations 1968 (SI 1968/1530)

Diving Operations at Work Regulations 1981 (SI 1981/399)

Work in Compressed Air Special Regulations 1958 (SI 1958/61), as amended by SI 1960/1307, SI 1973/36 and SI 1984/1593

Electrical Energy Regulations 1908 (SI 1908/1312), as amended by SI 1944/739

Woodworking Machines Regulations 1974 (SI 1974/903), as amended by SI 1978/1126

Protection of Eyes Regulations 1974 (SI 1974/1681), as amended by SI 1975/303

Control of Lead at Work Regulations 1980 (SI 1980/1248)

Ionising Radiations Regulations 1985 (SI 1985/1333), as amended by SI 1986/392

Asbestos: Products (Safety) Regulations 1985 (SI 1985/2042)

CONSTRUCTION OPERATIONS

Type of plant equipment or job involved	A Testing and thorough examination			B Thorough examination			C Inspection			D References
	Testing and thorough examination	Who carries out this work	Results to be recorded on Form No	Thorough examination	Who carries out this work	Results to be recorded on Form No	Inspection to be carried out	Who carries out this work	Results to be recorded on Form No	Legal reference
SCAFFOLDING							weekly or more often in bad weather	competent person	form 91 (pt1) A	W.P. regn 22
EXCAVATIONS EARTHWORKS TRENCHES SHAFTS AND TUNNELS				weekly or more often if part has been affected, e.g. explosives collapse	competent person	form 91 (pt I)B entry to be made day of examination	at least every day or at start of shift	competent person		G.P. regn 9
MATERIALS OR TIMBER USED TO CONSTRUCT OR SUPPORT TRENCHES EXCAVATIONS COFFER DAMS CAISSONS							on each occasion before use	competent person		G.P. regn 10 (1) G.P. regn 17 (2)
COFFER DAMS CAISSONS				before men are employed therein and at least weekly or more often if explosives are used or any part is damaged	competent person	form 91 (pt I)B	daily and before men are employed therein	competent person		G.P. regn 18
DANGEROUS ATMOSPHERES				before men are employed therein and as frequently as necessary	competent person. Instrument may be necessary	in any convenient way to show how examination was done				G.P. regn (21) (c)
CRANES (all types) CRABS WINCHES	once every four years and after substantial alteration or repair	competent person, normally by insurance co. engineer, manufacturer or erector	crane: form 96 crab: form 80 winch: form 80	at least every 14 months	competent person e.g. insurance co. engineer	form 91 (pt I) J or on a special filing card containing the prescribed particulars	weekly	competent person e.g. crane driver	form 91 (pt1) C–F	L.O. regn 10 (1) (c) L.O. regn 28 (1) (2) and (3)
PULLEY BLOCKS GIN WHEELS SHEER LEGS	before first use and after alteration or substantial repair unless used only for loads under 1 ton	competent person, normally the manufacturer or insurance co. engineer	form 80	at least every 14 months	competent person, e.g. insurance co. engineer	form 91 (pt II) J or on special filing card containing the prescribed particulars	weekly	competent person	form 91 (pt1) C–F	L.O. regn 10 (1) (2) L.O. regn 28 (1) and (2)

Figure 10.1

Type of plant equipment or job involved	**A** Testing and thorough examination			**B** Thorough examination			**C** Inspection			**D** References
	Testing and thorough examination	Who carries out this work	Results to be recorded on Form No	Thorough examination	Who carries out this work	Results to be recorded on Form No	Inspection to be carried out	Who carries out this work	Results to be recorded on Form No	Legal reference
CHAINS ROPE SLINGS AND LIFTING GEAR	before first use and after alterations or repair	competent person, normally manufacturer	form 97	at least every 6 months, except when used only occasionally	competent person, e.g. insurance co. engineer or at a testing house	form 91 (pt I) J or on a special filing card containing the prescribed particulars	SPECIAL NOTE: chains or lifting gear which have to be annealed, see form 91 (pt I) for detail and L.O. 41	chains of lifting gear which have to be annealed, see form 91 (pt I)	form 91 (pt I) or form 1946 containing the prescribed particulars	L.O. regn 34 L.O. regn 40 L.O. regn 41
WIRE ROPE	before first use	manufacturer	form 87	at least every 6 months except when used only occasionally	competent person, e.g. insurance co.	form 91 (pt I) J				L.O. regn 34
STEAM BOILER (new)	before use	manufacturer or boiler inspecting co.	no special form							F.A. 1961 s. 33
STEAM BOILER (cold)				every 14 months and after extensive repairs	competent person, e.g. insurance co. engineer	form 55				F.A. 1961 s. 33 (4)
STEAM BOILER UNDER PRESSURE				every 14 months and after extensive repairs	competent person, e.g. insurance co. engineer	form 55A				F.A. 1961 s. 33
STEAM RECEIVERS AND CONTAINERS	not required		a certificate as to the safe working pressure provided by maker	at least every 26 months	competent person, e.g. insurance co. engineer	form 58				F.A. 1961 s. 35
AIR RECEIVERS	before use	manufacturer or insurance co. engineer	a certificate as to the safe working pressure provided by maker	at least every 26 months (see special conditions)	competent person, e.g. insurance co. engineer	form 59				F.A. 1961 s. 36

NOTE:
re abbreviations used
F.A. 1961   Factories Act 1961
W.P. regn   Construction (Working Places) Regulations 1966
G.P. regn   Construction (General Provisions) Regulations 1961
L.O. regn   Construction (Lifting Operations) Regulations 1961
NOTE:
definition of lifting appliance
means a crab, winch, pulley-block or gin-wheel for raising or lowering and a hoist, crane, sheer-legs, excavator, dragline, piling-frame, aerial cableway, aerial ropeway or overhead runway.

definition of lifting gear
means a chain, sling, rope-sling or similar gear and a ring, hook plate, clamp, shackle, swivel and eye bolt.

The forms mentioned for result recording are those prescribed and are available from HMSO

competent person
There is no legal definition. The person who is selected or appointed to act as a competent person must have practical and theoretical knowledge together with actual experience on the type of plant, machinery, equipment or work which he is called upon to examine.
Such knowledge and experience will enable him to detect faults, weakness, defects, etc., which it is the purpose of the examination to discover and assess.

NOTE:
For other tests, examinations and inspections in connection with specialised operations, such as diving, work in compressed air, ionising radiations, etc., expert advice should be sought.

EXEMPTIONS:
Crawler-tracked shovel or dragline excavators
Such machines are occasionally used as cranes solely by the attachment of lifting gear to the shovel or bucket for work immediately connected with excavations the machine has been directly engaged on; this is only permissible provided a competent person specifies the maximum load or loads to be lifted. The maximum load or loads and the lengths of jib or boom to which they relate, together with a means of indentification, must be plainly marked upon the excavator. The Certificate—Form 2209—must be completed.

Legal Reference:
F.A. 1961: The Construction (Lifting Operations) Regulations 1961—Certificate of Exemption No. 2 (General).

Figure 10.1 (continued)

Type of plant equipment or job involved	A Testing and thorough examination			B Thorough examination			C Inspection			D References
	Testing and thorough examination work	Who carries out this work	Results to be recorded on Form No	Thorough examination	Who carries out this work	Results to be recorded on Form No	Inspection to be carried out	Who carries out this work	Results to be recorded on Form No	Legal reference
CRANES appliances for anchorage or ballasting				on each occasion before crane is erected	competent person, e.g. crane erector of fitter					L.O. regn 19 (3)
CRANES test of anchorage or ballasting	before crane is taken into use, i.e. after each erection or re-erection on a site or whenever anchorage or ballasting arrang. changed	competent person, normally crane erector in presence of insurance co. engineer	form 91 (pt1) D	has to be done after exposure of crane to weather conditions likely to have affected its stability. A re-test might be necessary	competent person, e.g. insurance co.					L.O. regn 19 (4)
CRANES test of automatic safe load indicator (jib cranes)	after erection or installation of crane and before it is taken into use	crane erector or insurance co. engineer; must be a competent person with knowledge of the working arrangements of indicator	form 91 (pt1) E				weekly	competent person, e.g. crane driver or fitter	form 91 (pt1) E NOTE: this will be part of normal weekly inspection	L.O. regn 30
CRANES mobile jib test of automatic safe load indicator	before crane is taken into use, after it has been dismantled or after anything has been done which is likely to affect the proper operation of indicator. e.g. change in jib length	competent person, e.g. erector, manufacturer, engineer, insurance co.	form 91 (pt1)				weekly	competent person, e.g. crane driver	form 91 (pt1) E NOTE: this will be part of normal weekly inspection	L.O. regn 30
LIFTING other **APPLICANCES** i.e. excavator dragline piling frame, aerial cableway or ropeway, overhead runway				at least every 14 months or after substantial alteration or repair	competent person	form 91 (pt1) G-K or on a special filing card containing the prescribed particulars	weekly	competent person, e.g. driver	form 91 (pt1) C	L.O. regn 28 L.O. regn 10
HOISTS (goods) made altered or repaired after 1st of March 1962	before first use and after substantial alteration or repair	competent person, manufacturer or insurance co. engineer	form 75	at least every 6 months	competent person, e.g. insurance co. engineer	form 91 (pt1) G-K or on special filing card containing the prescribed particulars	weekly	competent person, e.g. fitter	form 91 (pt1) H	L.O. regn 46
HOISTS passenger	before first use, after re-erection, alterations in height of travel after repair after alterations	competent person, e.g. manufacturer, insurance co. engineer or erector	form 75 or form 91 (pt1) F following alterations to height of travel	at least every 6 months	competent person, e.g. insurance co. engineer	form 91 (pt1) G-K or on special filing card containing the prescribed particulars	weekly	competent person, e.g. fitter	form 91 (pt1)	L.O. regn 46

Figure 10.1 (above, and on pages 160 and 161) Guide to statutory tests, examinations and inspections

Asbestos (Prohibitions) Regulations 1985 (SI 1985/910), as amended by SI 1988/711

Abrasive Wheels Regulations 1970 (SI 1970/535)

High Flammable Liquids and Liquefied Petroleum Gases Regulations 1972 (SI 1972/917)

Offices, Shops and Railway Premises Act 1963 (Repeals and Modifications) Regulations 1974 (SI 1974/1943), as amended by SI 1981/917.

Petroleum (Consolidation) Act 1928 (Enforcement) Regulations 1979 (SI 1979/427), as amended by SI 1986/1951

Explosives Acts 1875 and 1923 etc. (Repeals and Modifications) Regulations 1974 (SI 1974/1885), as amended by SI 1974/2166

Employment of Young Persons (Glass Containers) Regulations 1955 (SI 1955/274)

Employment of Young Persons (Iron and Steel Industry) Regulations 1959 (SI 1959/756)

Employment Medical Advisory Service (Factories Act Order etc. Amendment) Order 1973 (SI 1973/36), as amended by SI 1985/1333

Fire Certificates (Special Premises) Regulations 1976 (SI 1976/2003), as amended by SI 1985/1333

Safety Representatives and Safety Committees Regulations 1977 (SI 1977/500)

Control of Substances Hazardous to Health (COSHH) Regulations 1988

BIBLIOGRAPHY

COSHH Assessments (A Step-by-step Guide to Assessment and the Skills Needed for It), HMSO

Hazard and Risk Explained, leaflet available free from HSE

Health and Safety Executive. *Guidance Notes*. A series of guidance notes issued under five main headings: Medical (M.S.), Environmental Hygiene (E.H.), Chemical Safety (C.S.), Plant and Machinery (P.M.) and General (G.S.). Published by HMSO. Relevant editions:

G.S.2	*Metrication of Construction Safety Regulations*
G.S.6	*Avoidance of Danger from Overhead Electric Lines*
G.S.7	*Accidents to Children on Construction Sites*
P.M.3	*Erection and Dismantling of Tower Cranes*
P.M.9	*Access to Tower Cranes*
P.M.15	*Safety in the Use of Timber Pallets*
P.M.16	*Eyebolts*
P.M.17	*Pneumatic Nailing and Stapling Tools*
P.M.20	*Cable-laid Slings and Grommets*
P.M.21	*Safety in the Use of Woodworking Machines*

P.M.22   *Mounting of Abrasive Wheels*
E.H.10   *Asbestos*
M.S.13   *Asbestos*

Health and Safety Commission Report (1984/85)
Health and Safety Executive (1976) *Construction – Health and Safety*, HMSO
Health and Safety Executive (1977–78). *Construction – Health and Safety*, HMSO
Health and Safety Executive (1979–80). *Construction – Health and Safety*, HMSO
Health and Safety Executive (1980). *Management's Responsibilities in the Safe Operation of Mobile Cranes*, HMSO (A report on three crane accidents)
Health and Safety Executive (1981–82). *Construction – Health and Safety*, HMSO
Health and Safety Executive (1988). *Blackspot Construction*, HMSO (a study of 5 years' fatal accidents in the building and civil engineering industries)
Health and Safety at Work Act 1974, HMSO
H. M. Chief Inspector of Factories (1986/87). *Health and Safety at Work* (Health and Safety Executive) [Report]
Health and Safety at Work Series. A series of booklets produced by the Health and Safety Executive, published by HMSO, of which the following are relevant to construction work generally and to the use of construction equipment:

No. 6A *Safety in Construction Work*: *General Site Safety Practice*
No. 6B *Safety in Construction Work*: *Roofing*
No. 6C *Safety in Construction Work*: *Excavation*
No. 6D *Safety in Construction Work*: *Scaffolding*
No. 6E *Safety in Construction Work*: *Demolition*
No. 6F *Safety in Construction Work*: *System Building*

Introducing COSHH (A Brief Guide for All Employers), leaflet available free from HSE
Lomax, J. (1987). Safety and health in construction. *Safety Practitioner*, **5**, No. 9, 11–15
Ministry of Labour (1966). *Accidents in the Construction Industry*, HMSO
Nursing Times, Nursing Mirror (1986). Danger: men at work. **82**, No. 35, 44–45, 47–48.
The Control of Substances Hazardous to Health Regulations 1988, Approved Code of Practice Control of Substances Hazardous to Health

and Approved Code of Practice Control of Carcinogenic Substances, HMSO
Training Department, Richards and Wallington Industries Ltd (1979). *Tower Crane Tester's Handbook*

Chapter 11
Insurance and Licensing Legalities

11.1 INSURANCE

Construction companies and equipment hire companies require a variety of insurance policies and cover. The need for insurance cover could arise because (1) it may be required by law and is therefore compulsory; (2) it may be required by a contractual arrangement entered into by the company and other party; and (3) it is sensible to minimise the company's risks, although it may not be a requirement placed on the company either by law or by contract.

Examples of compulsory insurance are the insurance of vehicles as required by the Road Traffic Act 1972, the insurance against claims from employees as required by the Employers' Liability Act 1969, and the insurance against claims from the public as required by the Finance (No. 2) Act 1975.

Examples of insurances required by contractual agreements are the responsibilities of contractors under ICE Conditions of Contract or the JCT Standard Form of Building Contract and the insurance required by certain hiring agreements between equipment hire companies and the hirer.

Insurance precautions, of course, can go beyond satisfying legal or contractual requirements. The Road Traffic Act only requires a specified minimum insurance cover, but the owner of a vehicle may insure it comprehensively, to reduce his risks and ensure adequate compensation in the event of loss or accidental damage. Insurance of buildings and insurance of contents are other examples. Some hire agreements between equipment companies and contractors do not specify insurance but place the responsibility for loss or damage of the equipment with the hirer. In such cases insurance is not a contractual requirement but is clearly a sensible precaution.

The broad classes of insurance that are available are: (1) liability insurance, (2) material damage insurance, (3) pecuniary insurance and (4) benefit insurance.

Liability insurances, as far as contractors and equipment companies are concerned, are:

- Employers' liability
- Public liability
- Motor insurance, third party
- Liability under contract, such as ICE Conditions of Contract or JCT Standard Form of Contract.

Material damage insurance covers such items as:

- Insurance of works which may be specified in the contract between the employer (i.e. the promoter of the works) and the contractor
- Insurance of buildings and contents
- Insurance of plant and equipment
- Engineering insurance
- Motor insurance covering accidental damage.

Pecuniary insurances cover fidelity risks, credit risks and consequential loss.
Benefit insurance covers personal accident.

11.1.1   Employers' Liability Insurance

Employers' liability insurance is required by the Employers' Liability (Compulsory Insurance) Act 1969, which became effective from 1972. This Act requires a company, the employer or master, to take an insurance policy to cover the employer's liability to his employees for bodily injury or disease arising out of and in the course of their employment.

Employers' liability insurance is in addition to the responsibilities placed on the employer by the Health and Safety at Work Act as described in Chapter 10. Two issues of this insurance that are relevant to the construction industry are those of labour-only subcontractors and hired-in equipment. Labour-only subcontractors present a problem of definition as to whether the labourers are, for insurance purposes, to be treated as employees. The custom has been to regard them as such for the purposes of employers' liability insurance and to have the insurance extended to cover this.

The situation with regard to hired-in equipment with an operator is that while the owner of the equipment, the hiring company, remains the operator's employer, the contract of hire entered into between the owner and the hirer usually requires the hirer to accept liability for the operator (see Clause 8 of the Contractors Plant Association Conditions for Hiring Plant). Although the law requires the owner to protect his employee, the contract between the owner and the hirer gives the owner indemnity against claims by the operator which would normally be regarded as employer's liability. The custom has developed for the hirer to indemnify the owner, but if the hirer is not a company but a private individual excavating for his garage foundations, a different view would prevail and the hirer would not be required to carry this responsibility.

11.1.2   Public Liability Insurance

The Finance (No. 2) Act 1975 requires that payments to subcontractors can be made without deduction of tax only if the subcontractor has an exemption certificate from the Inland Revenue. One requirement in obtaining an exemption certificate is that the company must have an insurance policy covering public liability. This insurance is required to cover claims from the public for bodily injury or disease. No requirement exists to cover claims from the public with regard to property damage, although companies may insure themselves against this also and usually do. Usually a public liability policy will cover claims from the public for bodily injury and disease and for property damage.

The situation regarding hired-in equipment is similar to that of employers' liability, in that hire agreements between equipment companies and contractors (usually based on the Contractors' Plant Association Conditions for Hiring Plant) contain a clause requiring the hirer to indemnify the owner against claims from the public for injury or damage to property. The common forms of contract found in the construction industry usually require the contractor to indemnify the employer (i.e. the promoter of the works) against claims from the public.

11.1.3   Contractors' All-risk Policy (Mainly Material Damage Insurance)

A contractors' all-risk policy is usually entered into jointly by the promoter of the construction work and the contractor, and this policy provides the main protection for the works under construction. Usually the items protected against loss or damage in an all-risk policy are: the permanent and temporary construction works and all the materials connected with the works; plant, equipment, tools and temporary buildings; and employee's personal effects if not covered by other insurances.

The policy normally includes cover while materials or other items are in transit to and from the site. It is also possible to arrange inflation protection cover in respect of the additional cost of reconstruction after some mishap. The contractors' all-risks policy can be extended to include consequential loss as well as the material damage aspects listed above. The consequential loss cover would protect the promoter of the construction work against loss of income, such as rent from an incomplete building delayed by some mishap. The full value of construction equipment would normally be insured within a contractors' all-risk policy. The ICE Conditions of Contract require the construction plant to be insured but the JCT form of contract does not.

Hire agreements between equipment hire companies and contractors usually require the hirer to indemnify the owner of the equipment against loss or damage to the equipment. The cover given in a contractors' all-risk policy would normally include the full value of the equipment and therefore protect the hirer against claims from the owner of the equipment.

11.1.4 Plant and Equipment Insurance (Material Damage Insurance)

Contractors who own equipment and equipment hire companies will normally ensure that their plant and equipment is insured, even though there may be no legal or contractual requirement. Insurance of equipment can be arranged separately or can be arranged as extensions to other policies, mainly the contractors' all-risk policy or an engineering policy, which is dealt with in the next section, or as part of a commercial vehicle policy, if the equipment is mechanically propelled.

A contractor's equipment policy would be a material damage policy and would cover loss or damage while insured. The same cover could be obtained in a contractors' all-risk policy which also includes the travelling of equipment to and from the site. An engineering policy would normally be taken to protect against breakdown and as a means of acquiring an inspection service by a competent person (Chapter 9 also refers to this), but the policy may sometimes be extended to cover loss or damage. Mechanically propelled equipment such as dumpers, excavators, graders, etc., may be insured within a contractor's commercial vehicle policy.

In hire agreements between equipment hire companies (the owner) and contractors (the hirer) the responsibility for loss or damage of the hired equipment normally rests with the hirer. The hirer may be responsible for lost revenue after the hire period has expired while the equipment is being repaired or replaced. The hirer may not necessarily be required to insure the equipment but if he does not do so, he will be carrying the risk himself. The Contractors' Plant Association Conditions of Hiring Plant are an example whereby the hirer is required to make good to the owner all loss of or damage to the equipment from whatever cause, and in these circumstances the hirer would be well advised to insure against such claims from the owner. This insurance may be included under the policies described above.

11.1.5   Engineering Insurance (Material Damage Insurance, Breakdown and Inspection)

An engineering insurance policy can be taken to cover breakdown and accidental damage risks to lifting machinery, breakdown risks to electrical and mechanical equipment, and the risk of explosion to boilers and pressure equipment. Trucks, tractors and dumpers, etc., could be included in an engineering policy but would be more likely to be included in a motor insurance policy.

If lifting machinery is taken as the main example of interest to the construction industry, the two main risks are mechanical and/or electrical breakdown and accidental damage. Insurance against mechanical or electrical breakdown would include the breaking or burning out of part of the equipment, arising from a mechanical or electrical defect, causing the equipment to stop and requiring immediate repair or replacement. Frost damage would also be covered, but the main exclusion to such policies would be 'wear and tear'. Excavators can be insured under the same policies as cranes, and clauses for accidental damage will cover damage to the equipment from extraneous causes. The engineering policy can also include insurance against damage to property and to the goods being lifted.

The important feature of an engineering policy is that it includes an inspection service by a competent person. The insurance companies who specialise in engineering policies employ qualified engineers to undertake inspections. The inspection service can be acquired separately without an insurance policy and can be arranged to meet the requirements of the Health and Safety at Work Act (see Chapter 10). The competent person provided by the insurance company can certify that the inspections have taken place.

11.1.6   Motor Insurance (Liability and Material Damage Insurance)

The Road Traffic Act of 1972 requires certain minimum insurance for motor vehicles used on the road. A motor vehicle used on the roads must be insured to cover death or bodily injury arising out of the use of the vehicle. The liability is up to an unlimited amount and includes emergency treatment fees. The Act also requires that the policy holder must be issued with a certificate of insurance and that the driver involved in an accident must produce the certificate of insurance.

The motor insurances that are available extend the cover beyond the minimum required by the Road Traffic Act. The four main types are (1) comprehensive, (2) third party, fire and theft, (3) third party only and (4) Road Traffic Act only.

The comprehensive policy provides the greatest cover and includes third party liability for death, bodily injury, emergency treatment fees and damage to property, including other vehicles. The comprehensive policy also provides cover for loss or damage to the insured's vehicle, caused by accidental damage, fire or theft.

Third party, fire and theft policies cover the third party liability, as in comprehensive policies, and loss or damage to the insured's vehicle due to fire or theft. Accidental damage to the insured's vehicle, which is covered in a comprehensive policy, is excluded.

Third party only policies cover liabilities to third parties, as in comprehensive policies, but exclude loss or damage to the insured's own vehicle.

Road Traffic Act only policies meet the minimum requirements of the Road Traffic Act, which are third party liabilities for death or bodily injuries to third parties arising out of use of the vehicle on the road and fees for emergency treatment.

Construction Vehicles

Special vehicles or certain items of construction equipment can be insured under a commercial motor vehicle policy. Such vehicles or items of equipment can be grouped as follows: digging machines; site clearing and levelling equipment; mobile cranes; mobile equipment – e.g. compressors, welding and spraying equipment; dumpers and road rollers.

A goods carrying vehicle fitted with lifting equipment for the purposes of loading goods onto itself is treated as a goods carrying vehicle for insurance purposes. With respect to mobile cranes, the crane itself requires inspection services that are included in an engineering policy. An engineering policy would also provide for breakdown, accidental damage and third party liability, but excludes the liability protection required by the Road Traffic Act. Thus, there is a need for both motor insurance and an engineering insurance on, for example, a mobile crane, and there may be a risk of duplication if the insurance cover is not carefully arranged.

If the mobile crane is hired out, the motor insurance cover can be extended to include use while on hire. Dumpers can be included in a motor insurance policy and the cover can be extended to include use by a hirer. Mobile equipment such as excavators and shovels can be insured as part of a motor insurance policy and can include damage to the insured's own equipment. The cover can also be extended to the hirer. Protection against third party liabilities can be included, but the third party risk other than those arising under the Road Traffic Act may be covered better by a public liability policy.

Site clearing and levelling equipment is grouped separately from mobile equipment, as this class of equipment cannot excavate below the level of the wheel base. Third party risks to property in the ground, such as water, drainage and other services, are much reduced and therefore attract different premiums.

Contingent Liability

A motor contingent liability policy is a policy that protects the insured against liability to third parties resulting from the use of vehicles on behalf of the insured, over which the insured does not have immediate control.

As an example, suppose that an equipment hire company believes that the hirer's insurance has been extended to indemnify the owner but that the policy, by oversight, has not been extended, or has lapsed. A contingent liability policy would indemnify the owner in these circumstances. Similarly, a hirer may believe that the owner's motor policy extends its protection to the hirer, but if the owner's policy has lapsed, or is not extended to cover the particular circumstances, then the contingent liability policy could provide protection. This policy is important, because, under hire agreements based on the Contractors' Plant Association Conditions, it is the *hirer* who is responsible for third party liabilities such as injury or property damage caused by the plant. Thus, the *hirer* would bear the costs of such claims if the hire company's policy failed to operate.

11.1.7   Summary of Insurance and Equipment

This review of insurance policies has described the main areas of insurance cover required by the owners and operators of construction equipment. It has also described how insurance can be arranged through a variety of different policies, such as employers' liability insurance, public liability insurance, contractors' all-risk insurance, engineering insurance and motor insurance. The dangers to avoid in arranging insurance cover are duplication, which increases costs, and gaps in the cover, which leave the potential liability or loss uninsured. It is unlikely that such duplication or gaps will arise when equipment is owned and used by the same company, because all insurances will probably be arranged with the same company or broker and the insurance needs of the company can be arranged in a co-ordinated manner. The main dangers of duplication and gaps arise when equipment is hired out or hired in, and it is therefore necessary to check all the insurance arrangements in these circumstances.

All companies require employers' liability insurance and most require public liability insurance. A mobile crane would require motor insurance and engineering insurance to cover breakdowns and inspections. A crane used on site would probably be covered in part by the contractors' all-risk policy but would also require engineering insurance to cover breakdowns and to acquire the inspection service. A field excavator used within a site would probably be covered within a contractors' all-risk policy. Thus, the same policy could cover both an excavator owned by the contractor and an excavator which he hired in.

In general, hire agreements place the responsibility for loss and damage to equipment and the liabilities with the hirer.

11.2   LICENSING

The use of public roads within the UK is controlled by extensive legislation, broadly called the Road Transport Laws, which control the construction and use of vehicles on the public roads. Within this legislation is a system of licensing which controls the use of vehicles and their drivers on the public roads and collects taxes. The three main licensing systems of interest to the construction industry are as follows:

(1) *Vehicle excise licensing*, which requires a licence, for which duty is payable, to be in force for all vehicles used on the public roads, unless exempted.
(2) *Driver's licensing*, which requires every driver of a motor vehicle to hold a driving licence. A prescribed driving test is taken before a driving licence is issued.
(3) *Operator's licensing*, a system of goods vehicle licensing which exercises control over operators to ensure the proper use and roadworthiness of the vehicles and observance of the drivers' hours law.

In addition to these three licensing systems, there is *public service vehicle licensing*, which controls vehicles, operators and drivers of passenger carrying vehicles, such as buses and taxis.

The laws relating to road transport which embody the above licensing systems are mainly the following:

Road Traffic Act 1960, 1972, 1974
Transport Act 1962, 1968, 1978, 1980, 1981, 1982, 1985
Road Traffic (Drivers Ages and Hours of Work) Act 1976
Road Traffic (Driving Licences) Act 1983
European Communities Act 1972
Vehicle and Driving Licences Act 1969
Vehicles (Excise) Act 1971
Finance Acts 1971, 1982, 1983, 1984, 1985, 1987

From these and other Acts are derived many regulations, general orders and EEC directives, such as the Motor Vehicles (Construction and Use) Regulations and the Motor Vehicles (Authorisation of Special Types) General Order No. 1198, 1979.

The road transport laws also require all road users to have a minimum insurance, as described earlier. These laws also cover all aspects of road use, such as lighting of vehicles, speed limits, weight limits, securing loads, crane hooks, etc., and the implications of the Weights and Measures Act 1985 with respect to carrying sand, ballast and ready mix concrete, and the need for a conveyance note. Further explanation of these requirements can be obtained from the Acts listed above and the references given.

11.2.1   Vehicle Excise Licensing

General Requirements

A person who uses or keeps on a public road any mechanically propelled vehicle must have an excise licence in force, unless he is exempted. Excise licences are issued from local vehicle licensing offices and can also be renewed from post offices. A vehicle excise licence can be taken out for 6 months or 12 months. Applications for a licence or its renewal must be accompanied by a certificate of insurance and a test certificate if the vehicle is subject to a test procedure.

The annual duty for vehicles varies according to the class of vehicle as defined in the Vehicles (Excise) Act 1971. Among the classes defined are:

(1) Tractors for the purposes of agricultural and forestry work.
(2) Vehicles designed, constructed and used for trench digging or any kind of excavating or shovelling work, used on the public road for that purpose only or for proceeding to and from such but not carrying any load.

(3) Mobile cranes used on the public road only, either as cranes in connection with work being carried out on a site in the immediate vicinity or for the purpose of proceeding to and from a place where they are to be used as cranes, but not carrying any load.

(4) Works trucks, being goods vehicles designed for use in private premises and used on public roads for carrying goods between two premises or between premises and a vehicle, or in connection with road works.

(5) Haulage vehicles of different weight classes of up to 2 tons, 2–4 tons, 4–6 tons, 6–7¼ tons, 7¼–8 tons, 8–9 tons, 9–10 tons, 10–11 tons, over 11 tons.

(6) Goods vehicles. Taxation for goods vehicles is based on gross plated weight, and with heavier vehicles this is coupled with the number of axles. The Vehicle (Excise) Act 1971, Schedule 4 defines the classes; the Annual Finance Act defines the rate of licensing.

The various categories include:

- Small vehicles up to 1525 kg unladen.
- Vehicles up to 7.5 tonnes gross. This category includes other vehicles such as tower wagons and vehicles exempt from plating and testing.
- Vehicles of 7.5–12 tonnes gross
- Rigid vehicles over 12 tonnes gross. This category is divided into vehicles with 2, 3 and 4 axles.
- Drawbar trailers, 4–8 tonnes, 8–10 tonnes, 10–12 tonnes, 12–14 tonnes and over 14 tonnes. This category excludes gritters, snow ploughs, road construction vehicles, farming implements and trailer carrying or making gas to propel the drawing vehicle.
- Articulated vehicles, broken into 16 weight categories between 12 000 kg and 38 000 kg, and divided into 2-axle tractors and 3-axle tractors. Each tractor is then categorised with 1-, 2- or 3-axle trailers.
- Farmers' and showmen's goods vehicles.

Exemptions

Certain vehicles are exempt from paying duty under the Vehicles (Excise) Act 1971. Exemptions of particular interest to the construction industry include:

- Road construction vehicles used on a road to carry built-in road construction machinery.
- A vehicle which is to be used exclusively on roads not repairable at the public expense but details of the vehicle must be declared to the licensing authority.

- Subject to approval, no duty is payable on a vehicle which uses public roads only for passing from land in the owner's occupation to other land in his occupation for distances not exceeding 6 miles in any week. This is allowed by section 7(1) of the Vehicle (Excise) Act 1971.

Mobile Plant

Vehicles carrying no load other than built-in plant or machinery are taxed as goods vehicles. The weight of the built-in plant or machinery is deducted from the total weight in calculating the unladen weight of the vehicle for the purposes of assessing the duty payable.

11.2.2 Driver Licensing

No person may drive or permit another person to drive a motor vehicle on a road unless that person holds a driving licence granted under the conditions of the Road Traffic Act 1972.

The minimum age for drivers' licences is 16 to 21, depending on the class of vehicle. In the UK the minimum age is 17 for an agricultural tractor; for a medium-sized goods vehicle the minimum age is 18; and for larger vehicles it is 21. The minimum age for a goods vehicle not exceeding 7.5 metric tonnes is 18 within the EEC and 21 for heavier vehicles, although this can be reduced for drivers who have attended goods vehicle training courses and hold certificates of competence. The driver must pass a driving test, conducted by examiners appointed by the Licensing Authority, to obtain a driving licence. Schedule 4 of the Road Traffic Act 1972 sets out the offences that can lead to the driver's licence being endorsed or disqualified.

The classes of vehicle for which licences can be obtained are given in Table 11.1.

Heavy Goods Vehicles

A person must not drive nor permit another person to drive a vehicle classed as a heavy goods vehicle (HGV) unless the driver holds an HGV licence authorising him to drive vehicles of that class. A heavy goods vehicle is defined as a large goods vehicle which is constructed or adapted to carry goods, the permissible maximum weight of which exceeds 7.5 tonnes, or an articulated goods vehicle. The maximum weight of the goods vehicle or trailer usually will be stated as the maximum gross weight on the

Table 11.1 Groups of motor vehicles for driving tests

Group	Class of vehicle in group	Additional groups covered
A	A vehicle without automatic transmission of any class not included in any other group	B, C, E, F, K, L, N
B	A vehicle with automatic transmission of any class not included in any other group	E, F, K, L, N
C	Motor tricycle weighing not more than 450 kg unladen, but excluding any vehicle included in group E, J, K or L	E, K, L
D	Motor bicycle (with or without sidecar) but excluding any vehicle included in group E or K	C, E
E	Moped	–
F	Agricultural tractor, but excluding any vehicle included in group H	K
G	Road roller	–
H	Track laying vehicle steered by its tracks	–
J	Invalid carriage	–
K	Mowing machine or vehicle controlled by a pedestrian	–
L	Vehicle propelled by electrical power, but excluding any vehicle included in group D, E, J or K	K
N	Vehicle exempt from duty under section 7(1) of the Vehicles (Excise) Act 1971	–

Department of the Environment plate, if fitted, or the manufacturer's plate.

Drivers of certain classes or types of vehicle are exempted from requiring an HGV licence. Among the exemptions that are relevant to the construction industry are:

- Track laying vehicles.
- Road rollers.
- Road construction vehicles used or kept on the road solely for the conveyance of built-in road construction machinery.
- Engineering plans.
- Works trucks.
- Industrial tractors.
- Digging machines.

To obtain an HGV licence, a driver must pass a test conducted by a Department of the Environment examiner. The classes of vehicle for which HGV licences can be obtained are given in Table 11.2.

Table 11.2 Classes of heavy goods vehicles

Class	Definition	Additional classes
1	An articulated vehicle not with automatic transmission	1A, 2, 2A, 3, 3A
1A	An articulated vehicle with automatic transmission	2A, 3A
2	A heavy goods vehicle not with automatic transmission, other than an articulated vehicle, designed and constructed to have more than four wheels in contact with the road surface	2A, 3, 3A
2A	A heavy goods vehicle with automatic transmission, other than an articulated vehicle designed and constructed to have more than four wheels in contact with the road surface	3A
3	A heavy goods vehicle not with automatic transmission, other than an articulated vehicle, designed and constructed to have not more than four wheels in contact with the road surface	3A
3A	A heavy goods vehicle with automatic transmission, other than an articulated vehicle, designed and constructed to have not more than four wheels in contact with the road surface	–

11.2.3 Operators' Licensing

General Requirements

The Operators' Licensing system is a system of goods vehicle licensing introduced by the Transport Act 1968, 1982, 1985 and the Road Traffic Act 1974, and modified to comply with the EEC Directive 74/561. All goods vehicles exceeding 3.5 tonnes gross plated weight need to be covered by an operator's licence. Exemptions from an operator's licence are as follows:

(1) Agricultural machinery and trailers.
(2) Dual-purpose vehicles and trailers.
(3) Vehicles and new trailers using the road for less than 6 miles per week while moving between private premises.
(4) Passenger-carrying vehicles and trailers.
(5) Hearses.
(6) Police, Fire Brigade or Ambulance.
(7) Fire-fighting and rescue vehicles at mines.
(8) Uncompleted vehicles on test or trial.

(9) Vehicles with limited trade plates.
(10) Visiting forces vehicles.
(11) Vehicles used by HM United Kingdom forces.
(12) A trailer incidentally used in construction.
(13) Road rollers and trailers.
(14) RNLI and Coastguard vehicles.
(15) Vehicles with special fixed equipment.
(16) Local Authority special vehicles for road cleaning and gritting.
(17) Civil Defence vehicles.
(18) Steam-propelled vehicles.
(19) Tower wagons.
(20) Vehicles used solely on aerodromes.
(21) Electrically propelled vehicles.
(22) Showmen's vehicles.
(23) Vehicles first used before 1 January 1972 not over 1525 kg unladen plated between 3.5 tonnes and $3\frac{1}{2}$ tons.
(24) Vehicles used for weighing vehicles or maintaining weighbridges.
(25) Water, electricity, gas or telephone emergency vehicles.

Vehicles which must be used under an operator's licence include all goods vehicles belonging to the licence holder or in his possession by virtue of a hire purchase agreement, hire or loan — i.e. all vehicles used by the licence holder, not simply owned by him. An operator's licence is required for each operating centre — i.e. yard or base for vehicles, in different Traffic Areas controlled by Licensing Authorities. Only one licence is required per Traffic Area, irrespective of the number of operating centres in the Traffic Area.

Application for an operator's licence is made to the Licensing Authority. Application may be made for additional vehicles not yet acquired and, if authorised, this simplifies the procedures of adding additional vehicles to the operator's fleet. The information the Licensing Authority will require includes:

(1) The use to which the vehicle will be put.
(2) Arrangements for ensuring that drivers keep within permitted hours of work and keep proper records.
(3) Vehicle maintenance facilities.
(4) Details of any past activities in operating vehicles for trade purposes by the applicant(s).
(5) Convictions relating to operating vehicles during the preceding 5 years by the applicant(s).
(6) Financial resources of the applicant(s).
(7) The names of company directors and officers of the applicant company and any parent company or the names of any partners in a partnership.

The Licensing Authority, in deciding whether to grant a licence, will consider:

(1) Whether the operator is a fit person to hold a licence, bearing in mind past convictions relating to the roadworthiness of the operator's vehicles.
(2) Arrangements for ensuring that the law relating to drivers' hours and records will be complied with.
(3) Facilities for satisfactory maintenance.
(4) Arrangements for checking that vehicles are not overloaded.
(5) The suitability of the proposed operating centre.
(6) The financial resources for the proper operation of the business.
(7) The professional competence of the operator or his manager.

In deciding whether to grant a licence, the Licensing Authority may hold a public enquiry or take objections from a specified group of interested parties, including the police, the local authorities, certain trade associations and certain trade unions. A licence would normally be granted for 5 years, and additional vehicles may be applied for at the time of licence application or individually when acquired. Arrangements also exist for permanent one-for-one substitution of vehicles.

The Licensing Authority has disciplinary powers to curtail or revoke the licence if a material change in circumstances occurs, or for certain offences relating to roadworthiness of vehicles, drivers' hours and records, or plating and testing.

An annual fee is charged for each vehicle specified on the licence and a separate identity disc is issued for each vehicle on payment of the fee. The disc must be displayed on the windscreen of the vehicle.

Maintenance

An applicant for an operator's licence must satisfy the Licensing Authority that the maintenance facilities are such that the vehicles are kept in a safe and roadworthy condition at all times. Licensing Authorities expect more than the minimum routine maintenance specified by the manufacturers and more than the daily running checks made by drivers: they look for a convincing system of inspection and preventive maintenance. The Department of Transport *Goods Vehicle Tester's Manual* lists the inspection items for roadworthiness. Readers are advised to refer to this, but it can be said that a satisfactory preventive maintenance inspection system requires:

- Competent staff capable of recognising the significance of defects.
- A system of recording inspections, detailing what was inspected and the action taken, such as the remedial work done and who undertook the remedial work.

- Adequate facilities for such inspections, including means of under-vehicle inspection.
- A schedule of inspections the frequency of which is chosen to match the workload and work type of the vehicles.
- A drivers' reporting system whereby the driver can report vehicle defects.

The Licensing Authority requires that maintenance reports be kept for a minimum of 15 months. Even if operators contract out inspections and maintenance to service companies, they remain the users of the vehicle and are held responsible to the Licensing Authority for the condition of their vehicles.

Drivers' Hours

Drivers of goods vehicles are generally subject to EEC hours law, controlled by EEC Regulation 3820/85, but there are numerous exemptions. Drivers exempt from EEC rules are subject to the British hours law contained in Part VI of the Transport Act 1968. Journeys through countries not in the EEC are subject to a European Agreement on the Work of Crews of Vehicles Engaged in International Road Transport. This agreement is known as AETR. The AETR rules apply to the whole of any journey.

EEC Hours Law

These rules can be summarised briefly as follows:

(1) A week is from 00.00 hours Monday to 2400 hours Sunday.
(2) The driving period between any two daily rest periods must not exceed 9 hours. Twice a week it may be extended to 10 hours.
(3) After six daily driving periods a driver must take a weekly rest period. There are adjustments if the six daily driving periods do not take until the end of the sixth day.
(4) The total driving in any fortnight must not exceed 90 hours.
(5) After $4\frac{1}{2}$ hours a driver must have a break of 45 minutes. This can be replaced by breaks of 15 minutes spread over the driving period.
(6) In each 24 hour period a daily rest period of at least 11 consecutive hours must be taken; this may be reduced to 9 hours three times a week.
(7) During each week one daily rest period must be extended into a weekly rest period of 45 hours; this can be reduced to 36 hours at home base or 24 hours away from home base, but must be compensated for *en bloc* by the end of the third week in question.
(8) There must be no bonus payments for distance travelled or time taken.
(9) Delay relief is allowed to enable drivers to reach a suitable stopping place.

British Hours Law

A brief summary of these rules for drivers exempt from EEC rules is:

(1) A driver must not drive more than 10 hours and his working day must not exceed 11 hours.
(2) Off-road driving is subject to exemption.

AETR Rules

AETR rules apply to journeys outside EEC countries, and if the journey passes through EEC countries these rules apply to the whole journey. This international agreement is between EEC member states and Austria, Czechoslovakia, East Germany, Norway, Sweden, Yugoslavia and the USSR. The AETR rules are broadly similar to EEC rules as defined in Regulation 543/1969.

Drivers' Records

EEC Regulation 3821/85 requires a tachograph to be fitted and used for goods vehicles registered in a member state. The driver is responsible for returning the tachograph record sheets to his employer and producing records for the current week plus the last day of the previous week.

When tachographs are used, the keeping of a record book is not required. If a tachograph is not in use, a driver must be issued with a record book for the purpose of recording the daily and weekly working records, which are returned to the employer.

Plating and Testing

Heavy goods vehicles require a first examination at a Department of the Environment goods vehicle testing station not more than 12 months after first being registered. The first examination includes assessing the vehicle's axle and gross weights. These weights are recorded on a plate, which must be prominently displayed in the cab of the vehicle: hence, these weights are known as the 'plated weights'. The assessment of the weights is followed by a test of roadworthiness, and, if satisfactory, the vehicle is issued with a plating certificate and a roadworthiness test certificate. The roadworthiness test must be repeated every 12 months.

Light goods vehicles are subject to a first roadworthiness test after 3 years and every 12 months thereafter.

11.2.4 Construction and Use of Vehicles

The law governing the construction and use of vehicles on the road is the Road Vehicles (Construction and Use) Regulations 1978, 1986, 1987. If some vehicles do not conform to these regulations, its use can be authorised under the Motor Vehicles (Authorisation of Special Types) General Order 1979, amended 1981, 1984, 1985, 1986, 1987, 1989. This General Order is important to the construction industry in that it covers the movement of abnormal indivisible loads, wide loads and engineering plant, such as mobile cranes and outsized dumpers.

BIBLIOGRAPHY

Abrahamson, Max W. (1983). *Engineering Law and the ICE Contracts*, Applied Science (reprint)

Department of Transport (1980). *New Arrangements for Testing Heavy Goods Vehicles and Public Service Vehicles: A Policy Document*, HMSO

Department of Transport (1980). *Tachographs: A Question and Answer Guide*, HMSO

Department of Transport (1980). *Tachographs: EEC legislation on the Introduction of Recording Equipment in Road Transport*, HMSO. A consolidated version of Regulations (EEC) Nos. 1463/70, 1787/73 and 2828/77

Department of Transport (1980). *Vehicle Excise Duty. 'Tax on Possession'*, HMSO

Department of Transport (1983). *Vehicle Testing: A Guide to the Operation of the MOT Test*, HMSO

Department of Transport (1984). *Goods Vehicle Testers' Manual*, HMSO

Department of Transport (1984). *Safety of Loads on Vehicles*, HMSO

Department of Transport (1988) *Report/Road Traffic Law Review*, HMSO

Department of Transport (1988). *Vehicle Registration and Licensing*, HMSO

Duckworth, J. (1988). *Kitchin's Road Transport Law*, Butterworths

Eagleston, F. N. (1985). *Insurance under the ICE Contract*, George Godwin

Joint Contracts Tribunal (1980). *JCT Guide to the Standard Form of Building Contract*, 1980 edition

Joint Contracts Tribunal (1984). *Form of Contract for Works of Simple Content*

Lowe, D. (1989). *The Transport Manager's Operator's Handbook*, Kogan Page

National Federation of Building Trades Employers (1982). *Construction Safety*

Turner, D. F. (1984). *Standard Contracts for Building*, George Godwin

Section 4

Financial and Budgetary Control

Chapter 12
Budgetary Control and Costing

12.1 INTRODUCTION

A budget acts as a standard of measure against which actual performance may be compared. The Institute of Cost and Management Accountants defines budgetary control as 'the establishment of budgets, relating the responsibility of executives to the requirements of a policy and the continuous comparison of actual with budgeted results, either to secure by individual action the objective of that policy, or to provide a basis for its revision'. Budgetary control therefore involves:

- Setting targets.
- Monitoring progress.
- Taking corrective action when necessary.

Within a system of budgetary control, budgets are established to relate the financial requirements for the component parts of the firm over the forthcoming period of 12 months to the overall policy of the company. Budgets may be forward estimates of costs or revenues, and as such are usually derived from records of past performance adjusted for future expectations.

12.2 PREPARATION OF BUDGETS

The budgetary system comprises many individual budgets which are ultimately integrated into a master budget. The master budget (Table 12.4) is similar in form to a profit and loss account but, unlike the latter, it is based on forward estimates of costs and revenues and is therefore only a forecast of the anticipated profit to be earned. From such an estimate other factors related to future expectations may be projected, such as the rate of return on capital employed, dividends to shareholders, capital to be retained in the business for reinvestment in assets and similar items related to profitability.

At the start of any attempt to prepare budgets for business activity during the coming year, it is necessary to prepare a budget for the investments to be made in plant and equipment, since it is through these assets that a plant company's revenues and costs are generated. Subsequently, a sales and an operating budget may be synthesised and the cash flow requirements determined. By a gradual process budgets may be adjusted to keep within the constraints on financial resources available to the company, to culminate finally in the master budget, as shown in Figure 12.1.

12.3   TYPES OF BUDGET

The *capital investment* budget is determined by the availability of capital to the company and is thus a schedule of available loan capital, equity, leasing

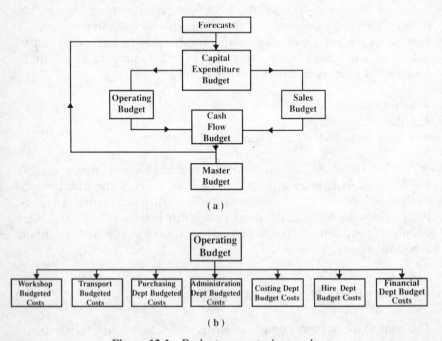

Figure 12.1   Budgetary control procedure

and retained profits. These sources of funds are treated more fully in Chapter 14, and the preparation of such a budget is covered in detail in Chapters 4 and 5.

The *cash flow* budget shows the short-term cash available period by period and is broadly determined by subtracting costs from revenues after taking into account payment delays. The preparation of a cash flow budget is described in detail in Chapter 13.

The *sales budget* for a rental or hire firm is simply the forecast of revenues from equipment hire. This will be made up of the expected income from the hire of individual items, which may fluctuate on a seasonal basis. Consequently, the budget should be prepared showing the anticipated annual and weekly incomes.

The *operating budget* is prepared from estimated costs of the planned requirements for materials such as fuel, lubricants, spare parts; staff and

Table 12.1   Workshop budget

Code (W)	Item	Annual (£)	Weekly (£)
Direct repair material and stock of spares			
10.0	Crawler cranes	6 000	120
20.0	Tower cranes	8 000	160
30.0	Trucks	4 000	80
40.0	Bulldozers	5 000	100
50.0	Loaders	7 000	140
60.0	Excavators	10 000	200
Cost		40 000	800
Direct labour costs			
10.1	Crawler cranes	7 000	140
20.1	Tower cranes	7 000	140
30.1	Trucks	5 000	100
40.1	Bulldozers	5 000	100
50.1	Loaders	8 000	160
60.1	Excavators	8 000	260
Cost		40 000	800
Indirect costs			
90.1	Storeman	6 000	120
90.2	Staff salaries	8 000	160
90.3	Rent	3 000	60
90.4	Rates	2 000	40
90.5	Electricity	500	10
90.6	Depreciation of tools and equipment	500	10
Cost		20 000	400
Total cost		100 000	2 000

Note: In practice the indirect costs may be separated into variable and fixed costs.

Table 12.2   Transport budget

Code (T)	Item	Annual (£)	Weekly (£)
Direct material costs			
100.0	Vans	4 000	80
110.0	8-ton lorries	5 000	100
120.0	16-ton lorries	5 000	100
130.0	32-ton lorries	3 000	60
140.0	Cars	3 000	60
Cost		20 000	400
Direct labour costs			
100.1	Vans	3 000	60
110.1	8-ton lorries	6 000	120
120.1	16-ton lorries	5 000	100
130.1	32-ton lorries	4 000	80
140.1	Cars	2 000	40
Cost		20 000	400
Indirect costs			
190.1	Staff salaries	11 000	200
190.2	Electricity	500	10
190.3	Rates	1 000	20
190.4	Rent	1 000	20
190.5	Vehicle depreciation	2 000	40
190.6	Tools and equipment	100	2
190.7	Telephone, etc.	400	8
Cost		15 000	300
Total cost		55 000	1 100

labour; equipment such as small tools, and depreciation of workshop equipment; and the business facilities, rent, rates, electricity, etc.

The difference in value between the sales and operating budgets is the anticipated profit before deduction of depreciation of the firm's assets of equipment and buildings, etc., for the year ahead.

The operating budget is subsequently subdivided into separate functions (see Figure 12.1b). These will include budgets for departments such as transport, workshop and administration. In larger concerns the latter may be broken down further into hiring, sales, costing, accounts and purchasing. By providing each department with a separate financial budget, a target is available against which subsequent performance may be monitored. Examples of the form of the annual budgets for the workshop, transport and administration functions are shown in Tables 12.1–12.3. It can be seen that the budgets consist of cost forecasts of the requirements for materials, labour and expenses. A coding system is used to allocate the

Table 12.3   Administration budget

Code (A)	Item	Annual (£)	Weekly (£)
Direct employment costs			
300.0	Staff salaries	9 500	190
Direct material costs			
310.0	Stationery	200	4
Direct expenses			
320.0	Photocopying	100	2
Indirect costs			
330.1	Telephone	1 000	20
330.2	Postage	700	14
330.3	Electricity	1 000	20
340.1	Rates	500	10
340.2	Rent	1 000	20
340.3	Office equipment	400	8
340.4	Insurances	600	12
Total cost		15 000	300

Table 12.4   Master budget

Code	Item	Annual (£)	Weekly (£)
Budgeted sales			
(R)	Plant hire	400 000	8 000
Total budgeted sales		400 000	8 000
Budgeted costs			
(W)	Workshop dept costs	100 000	2 000
(T)	Transport dept costs	55 000	1 100
(H)	Hire dept costs	20 000	400
(C)	Costing dept costs	25 000	500
(A)	Administration dept costs	15 000	300
(B)	Buying dept costs	25 000	500
(F)	Accounts dept costs	20 000	400
Total budgeted costs		260 000	5 200
Budgeted trading profit (sales minus costs)		140 000	2 800
Budgeted depreciation on plant and premises		100 000	2 000
Budgeted net profit before interest and tax		40 000	800

resources to particular departments or functions concerned, and like items
are collected under the same alphanumeric code. It is usual to present both
an annual budget and a weekly budget, as short-term fluctuations may be
the more usual pattern of expected performance.

12.4   CLASSIFICATION OF COSTS

For the preparation and monitoring of budgets, costs are collected and classified into the major functions: hiring, purchasing, workshop, transport, costing, administration/personnel, accounting. Within these functions the costs may be further collected into cost centres. For example, each item of equipment maintained by the workshop may be given a code number which may represent the cost centre of one or more similar items. The costs recorded for each cost function or centre may be subdivided into elements such as materials, labour and expenses:

(1) Material cost: consumables and spares.
(2) Labour cost: wages and salaries of the employees.
(3) Expenses: depreciation of plant and equipment, repairs, administration, services provided, water and electricity.

The costs of materials, labour and expenses which can be clearly allocated to a cost centre are called direct costs, and usually vary with the volume of production. Indirect costs are those materials, labour and expenses which cannot be directly identified to the cost centre, but which provide some function or service, such as a computer or the rent of the firm's offices and works. Indirect costs are thus apportioned between the cost centres, and are usually referred to as overhead costs.

Indirect costs are mostly fixed costs such as staff salaries, rent and rates, insurances, office equipment, maintenance tools and machines, which remain constant irrespective of the volume of work done. A direct or variable overhead is one which varies in cost with the volume of production, such as electricity.

12.5   COSTING

While budgets are prepared from predetermined costs, because of short-term changes in company performance it is essential that the actual costs incurred be continuously monitored and compared with budgeted costs in order that changes may be implemented. The difference between the actual cost and the predetermined cost is called a variance. A costing system should be updated regularly on a weekly basis, and the variances should be calculated for each function, department or cost centre. The procedure may also include analysis of the variances incurred by the individual items of plant in the fleet.

A note of caution with regard to the budgets, particularly departmental budgets, is advised. It is useful to compare the actual result with the value for the same month or week of the previous year. Astute managers can be adept at 'hiding' behind a 'stuffed' budget.

A budget may be 'stuffed' because of changed circumstances, e.g. reduction in the assumed rate of inflation, manipulation of the figures, etc. Managers' performance should therefore be measured against both the budget and previous year results.

12.5.1   Control of the Workshop Budget

EXAMPLE 12.1

The annual budget for a plant workshop is £100 000 (see Table 12.1). This figure is based upon the size of the fleet and the estimated hours that the fleet will be operated during the year.

The budgeted direct costs of the department consist of the purchase of consumable materials and spare parts and labour such as fitters, mechanics, etc., and these total £80 000. Budgeted indirect costs include staff salaries, rent, rates, insurances, general administration charges, depreciation of workshop equipment and power. These amount to £20 000 and are a fixed charge.

At the end of the year the firm's business activity had been lower than expected and the hours operated by the fleet were 10% fewer than initial estimates. The actual direct costs of the workshop over this period were £70 000 and the actual cost of overheads was £21 000. The position was:

	Planned budgeted costs for this date (£)	Adjusted budgeted costs of work done (£)	Actual cost (£)	Variance (£)
Direct costs	80 000	72 000	70 000	+ 2 000
Overheads	20 000	20 000	21 000	− 1 000
Total	100 000	92 000	91 000	+ 1 000

Thus, although the level of activity anticipated at the beginning had not been realised, the works department had managed to maintain a favourable variance on direct costs. In this example only overheads produced an unfavourable variance and a more detailed analysis of costs should reveal the reasons, such as excess secretarial staff. However, because the volume of business had not reached the level anticipated, the *budgeted profit* for the company also would not be fully recovered and the costs of the fixed overheads would have to be met from the reduced profits. For a plant hire company such consequences could be particularly severe, as much of the business costs are generated as fixed overheads. For a comprehensive review of performance, therefore, the sales variance should be included in the analysis.

12.5.2 Sales Variance

EXAMPLE 12.2

The budgeted hire revenue for a plant division over the coming 12 months is £400 000. Profit, overheads and depreciation are set at 10%, 20%, 20%, respectively, of revenue. However, the actual revenue was only £350 000. The variances recorded for direct and indirect costs (overheads) were, respectively, + £3 000 and + £1 000. In addition, several items were written off prematurely, leading to a negative variance on equipment depreciation of £1 000.

Analysis of Variance

Sales variance	= − £25 000	(i.e. 50% of £400 000 − £350 000)
Direct costs variance	= + £ 3 000	
Indirect costs variance	= + £ 1 000	
Depreciation variance	= − £ 1 000	
Actual shortfall on profits	£22 000	(i.e. £18 000 profit as compared to £40 000 budgeted profit)

It can be seen that the profit and overhead is under-recovered by £25 000 and only the collective economies made by the various departments reduced the magnitude of the shortfall to £22 000.

12.5.3 Other Useful Management Information

Equipment Utilisation Reports

The sales variance may also be calculated for each group or item of equipment, to provide management with a regular update of the effects of changes in the levels of utilisation and hire rates. The utilisation report is produced monthly and is divided into two sections, representing equipment for hire and non-operated plant — i.e. equipment hired without an operator. Each division is subdivided into individual cost centres or like machines — e.g. scrapers, bulldozers, crawler cranes, etc. — and the utilisation and price variance *on sales* for each item in a cost centre is produced in the final report as shown below.

Analysis of Variance

(1) *Utilisation variance* is the financial effect of using the equipment either more or less than those hours budgeted. Thus:

Utilisation variance = (actual hours × budgeted hire rate) −
(budgeted hours × budgeted hire rate)

(2) *Price variance* is the financial effect of charging more or less than the budgeted hire rate. Thus:

Price variance = (actual hours × actual hire rate) −
(actual hours × budgeted hire rate)

The sum of the utilisation and price variances multiplied by the profit and overhead margin is the true sales variance on that item.

EXAMPLE 12.3

During the month an item of equipment was hired out for 280 hours at a hire rate of £12 per hour. The budget anticipated only 200 hours of work at a hire rate of £13 per hour. Calculate the utilisation and price variances. The profit (and overhead) is set at 10% of the hire rate.

(1) Utilisation variance on sales
= (280 × 13) − (200 × 13) = 80 × 13 + £1 040

(favourable)

(2) Price variance on sales
= (280 × 12) − (280 × 13) = 280 × − 1 = − £280

(unfavourable)

Sales variance = (actual revenue − budgeted revenue) × profit and overhead margin
= (280 × 12 − 200 × 13) × 0.1 = + £760.

The effect of operating the equipment 80 hours more than planned increased the revenue by £1 040, reduced to £760 because the hire rate was less favourable than budgeted. The variances should be cumulatively totalled for each month, to present a comprehensive record, which, together with records collected of the hours operated for the machine, provides an indication of the competitiveness and excessive use or otherwise of the item.

These two variances signify to management the popular and competitive equipment items to operate, and the information is therefore valuable in deciding upon purchases and disposals.

Equipment Cost Report

A positive utilisation variance would usually be associated with increased maintenance, affecting direct costs and possibly indirect costs also. Therefore, the budgetary and cost control system recommended for departments should be installed for individual equipment items, and actual costs and revenues should be recorded on a regular basis. In this way all the variances may be monitored and the consequences on profitability quickly recognised.

EXAMPLE 12.4

During a 6 month period £4 000 revenue was received for a piece of equipment. The budget anticipated £5 000. Direct costs and indirect costs (overheads) from operating the various company departments were budgeted to the item at £1 500 and £2 000 respectively, with profit at 10% of turnover, and £1 000 was provided for depreciation. Actual direct costs recorded during the period were £1 250 and the actual overhead incurred was £1 950. Calculate the variances and percentage return on sales.

Analysis of Variance

Budgeted profit (10% of £5 000) $=$ £   500
Sales variance (£1 000 × 70%) $= -$ £   700
Direct cost variance

$$\left(£1\ 500 \times \frac{4\ 000}{5\ 000} - £1\ 250 \right) \qquad = -£ \quad 50$$

Overhead variance $= +£ \quad 50$
Depreciation variance $=$ £   0

Total variance $= -$ £   700

Actual profit on turnover $= \dfrac{-200}{4\ 000} = -5.0\%$ (loss)

Note: Sales variance multiplier of 70% is derived from profit, overhead and depreciation — i.e. (£500 + £2 000 + £1 000)/£5 000.

A comprehensive record of costs, current written-down value, revenue and profitability for each item may be used in conjunction with the utilisation report, to purchase and dispose of items at prices which are economic for the company. For many firms this type of information is monitored on an asset register of the equipment holdings.

12.6   MARGINAL COSTING

Marginal costs are those costs arising directly from the production process, which for a plant hire company would be largely those costs connected with maintenance and servicing of the equipment. They therefore vary directly with the hiring activity. Fixed costs, arising from the establishment charges, fluctuate very little with hiring levels. The purpose of the marginal costing method is to calculate the contribution made by each item of equipment for hire towards the fixed costs and profit of the business.

EXAMPLE 12.5

	Plant item, £00s, (weekly)				
	A	B	C	D	E
Hire revenue	10	9	5	15	10
Labour costs $\Big\}$ marginal costs	3	2	2	5	2
Material costs	3	2	1	5	3
Expenses	2	2	1	3	2
Contribution*	2	3	1	2	3
Contribution per £1 revenue	20p	33p	20p	13p	30p

* Contribution = overheads + profit.

This technique can be used to advantage during a short-term period when the market demand is low and hire rates need to be keen to attract custom, the contention being that any hire rate revenue which exceeds the marginal costs makes a contribution towards the fixed costs. However, such a pricing policy should be considered only during a short and difficult period, since the endeavour must be to realise the budgeted profit for each item over the 12 months periods. Thus, for plant item C, for example, a hire rate exceeding £400 per week will contribute to the fixed overhead, which may be a better alternative than leaving the machine idle. Conversely, the method gives a clear indication that the firm should be directing its sales effort on items B and E, as these machines can obtain favourable hire rates and give the best contribution towards fixed costs.

BIBLIOGRAPHY

Bathurst, P. E. and Butler, D. A. (1980). *Building Cost Control Techniques and Economics*, Heinemann

Clough, R. H. and Sears, G. A. (1979). *Construction Project Management*, Wiley

Cooke, B. and Jepson, W. D. (1979). *Cost and Financial Control of Construction*, Macmillan

Glautier, M. W. E. and Underdown, B. (1988). *Cost Accounting*, Pitman

Gobourne, J. (1982). *Cost Control in the Construction Industry*, Newnes-Butterworth

Izhar, R. (1988). *Costing and Management Accounting*, Oxford University Press

Merrit, A. J. and Sykes, D. (1965). *The Finance and Analysis of Capital Projects*, Longman

Pizzey, A. (1989). *Cost and Management Accounting*, Chapman and Hall

Sizer, J. (1985). *An Insight into Management Accounting*, Penguin

Chapter 13
Cash Flow

13.1 INTRODUCTION

Bankruptcy or voluntary liquidation is caused not only by a lack of cash, but also by an inability to raise cash in the form of loans or credit to meet immediate commitments, because creditors, investors and possible lenders of money — usually the banks — have lost confidence in the business and are not convinced that the company can continue to trade in a profitable and viable way. This loss of confidence is important because it is the existence of such confidence that permits overdrafts to be obtained and normal trade credit to be received, and a loss in confidence would result in existing creditors pressing harder for payment. Trade credit is an important factor in determining most companies' short-term cash requirements, and should trade credit be withheld, the short-term cash requirements increase significantly. The withholding of trade credit simply means that the suppliers to a company demand cash on delivery rather than invoicing, say, at the end of the month and requiring payment by 1 month later.

Most of Chapter 5 was concerned with profitability measured in terms of return on capital, and much of Chapter 14, dealing with financial management, will be concerned with determining the company's profit and distributing that profit. Profit seen simply as the difference between revenue and cost is a common measure of a company's well-being, and the derived ratios of profit/turnover and profit/capital employed are useful indicators of the company's performance, but these are derived from measuring profit. Undoubtedly, a company with a good profit is likely also to have good profitability (i.e. profit as a proportion of capital employed) and, in turn, be in a good state to avoid liquidity problems. But the company's liquidity needs to be monitored and managed also. Although a company may be profitable, it may have liquidity difficulties. An unexpected demand for payment may not be able to be met and may cause significant difficulties, if not bankruptcy.

The more detailed monitoring and managing of a company's cash flow can be seen as two related and integrated but different aspects. One is the cash required for normal trading operations and the other is the cash required for acquisitions, less disposals.

The cash required for normal trading operations is controlled by sales fluctuations, trade credit (creditors, less debtors), stocks, work in progress and perhaps value added tax. In equipment hire, sales fluctuations are manifest by the utilisation of equipment. Equipment is idle in a sales slump, but all equipment is highly utilised and more new equipment is required in a sales boom. Stocks in equipment hire are less important than in manufacturing and normally only represent spares, repair materials, fuel and oil. Also, equipment hire companies differ from manufacturing companies in that there is no manufacturing process absorbing manpower and materials and locking up cash. The nearest equivalent to 'work in progress' in equipment hire is equipment on hire for which the invoices have not been issued, or perhaps equipment under repair.

The other aspect is the cash required for the provision of the company's capital assets. This is particularly important to equipment hire companies, since as much as 50% of sales turnover may be used in meeting the costs of asset ownership. The variables that control this are the purchase and disposal of capital assets, the method of acquisition which, in turn, controls the methods of payment (e.g. purchase, hire purchase, or lease) and the company's ability to utilise capital allowances. In addition to these factors, the company's cash flow is also affected by interest and other bank charges, corporation tax and dividends. Corporation tax is important to the cash flow, as at the time of writing, capital allowances of 25% of the purchase price written down are allowed in acquiring an asset, such as an item of construction equipment. Therefore, the disposal and acquisition of equipment is significant when determining the corporation tax due and, in turn, the acquisition of equipment and payment of corporation tax are significant to the company's cash requirements. Dividends are also significant to the company's cash position, but within the control of the company's directors.

The following description of cash flow problems reflects the two main aspects of cash flow: (1) the cash flow resulting from the normal month-to-month trading operations and (2) the cash flows resulting from the acquisition and disposal of capital assets. These will be considered separately.

13.2   CASH FLOWS FROM NORMAL TRADING OPERATIONS

13.2.1   Trade Credit, Sales Fluctuations and Stocks

The factors that affect the short-term cash requirements in equipment hire are trade credit, sales fluctuations and stocks. The effects of these are best illustrated by considering an example which illustrates the effects of these factors one at a time. In manufacturing, the additional factor of 'work in progress' is important, but as there is no process in equipment hire that can be classed as work in progress, it is not significant and is not included in this example.

The monthly net cash flow stated in Example 13.1 is the difference between the cash outgoings and incomings in that month. The monthly contribution is the revenue that will be derived from the sales (hired-out equipment) during the month, less the direct costs in supporting the sales for labour, spare parts and consumables. The word 'contribution' is used because this money will be used to meet overhead costs and ownership costs and therefore cannot yet be described as 'profit'. All transactions in this example are exclusive of value added tax, which is referred to separately.

EXAMPLE 13.1

An equipment hire company initially holds 5 items of equipment which are hired out at the rate of £1 000 per month. Each month's maintenance on each item of equipment uses £200 in spare parts and £100 in consumables such as fuel, oil and grease. The fuel costs for running the equipment are the responsibility of the hirer. The workshop providing the maintenance support has sufficient labour to maintain 5 items of equipment in working order each month. The labour cost of the workshop is £750 per month. If all the 5 items of equipment were on hire and all the transactions were in cash, the monthly cash flow would be as follows:

Month 1 (with all transactions in cash)

	Cash out (£)	Cash in (£)
Sales (5 items of equipment on hire)		5 000
Workshop labour	750	
Purchases:		
Spares for 5 items of equipment for 1 month	1 000	

Consumables for 5 items of equipment for 1 month	500	
Totals	2 250	5 000

Net cash flow for month 1	+ £2 750
Contribution to company for month 1	+ £2 750

The contribution is defined as the revenue derived from the hire sales, less the direct costs incurred in supporting these sales, such as labour, spares and consumables. The company profit will be the contribution, less the ownership and overhead costs. If the hirers were given trade credit of 1 month to pay, then the cash flow for month 1 would be as follows.

Month 1 (with trade credit for hirers)

	Cash out (£)	Cash in (£)
Sales (5 items of equipment on hire)		nil
Workshop labour	750	
Purchases:		
Spares for 5 items of equipment for 1 month	1 000	
Consumables for 5 items of equipment for 1 month	500	
Totals	2 250	nil

Net cash flow for month 1	− £2 250
Contribution to company for month 1	+ £2 750

The company has the same contribution but the effect of trade credit to the hirers produces a net cash flow of − £2 250. If the suppliers of spares and consumables also offered 1 month trade credit but the labour was paid weekly, the cash flows would be as follows.

Month 1 (with Trade Credit for Hirers and from Suppliers)

	Cash out (£)	Cash in (£)
Sales (5 items of equipment on hire)		nil
Workshop labour	750	
Purchases:		
Spares for 5 items of equipment for 1 month	nil	
Consumables for 5 items of equipment for 1 month	nil	
Totals	750	nil

Net cash flow for month 1   − £ 750
Contribution to company for month 1   + £2 750

Thus, although the contribution generated in this month was again £2 750, the cash resources required were reduced to − £750 by the availability of trade credit from suppliers. Assuming that this is the normal trading experience, the cash flows in the following month, month 2, will be different from those in month 1, even though the same level of sales (i.e. hired-out equipment) is achieved. This is because the cash flows from month 1 sales and purchases will be present.

Month 2 (with Trade Credit for Hirers and from Suppliers)

	Cash out (£)	Cash in (£)
Sales (5 items of equipment on hire)		nil
Workshop labour	750	
Purchases:		
Spares for 5 items of equipment		
for 1 month	nil	
Consumables for 5 items of		
equipment for 1 month	nil	
Revenue from previous month's		
sales		5 000
Payments for previous month's		
purchases:		
Spares	1 000	
Consumables	500	
Totals	2 250	5 000

Net cash flow for month 2   + £2 750
Contribution to company for month 2   + £2 750

In month 2 the cash flow is a net inflow of +£2 750, arising from the revenue from month 1 sales. The contribution earned in month 2 was again £2 750 and the cash flows and contributions for the two months are:

	Cash flow (£)	Contribution (£)
Month 1	− 750	+ 2 750
Month 2	+ 2 750	+ 2 750

Provided that the 5 items of equipment are hired out and maintained each month, the cash flows and contributions will continue as in month 2. If in months 3 and 4 only 2 items of equipment are hired out, the pattern is disturbed, as follows.

Month 3 (with Trade Credit for Hirers and from Suppliers)

	Cash out (£)	Cash in (£)
Sales (2 items of equipment on hire)		nil
Workshop labour	750	
Purchases:		
Spares for 2 items of equipment for 1 month	nil	
Consumables for 2 items of equipment for 1 month	nil	
Revenue from previous month's sales		5 000
Payments for previous month's purchases:		
Spares	1 000	
Consumables	500	
Totals	2 250	5 000

Net cash flow for month 3   + £2 750
Contribution to company for month 3   + £ 650
Number of idle items of equipment   3
(The contribution is calculated as the revenue from 2 items of equipment of £2 000, less the workshop labour of £750, and the spares for 2 items of equipment of £400 and consumables for 2 items of equipment of £200.)

Although the sales in month 3 were reduced, the effect was not shown immediately on the cash flow because of trade credit. The reduction in contribution generated in the month is aggravated by the fact that the labour costs could not be reduced in a similar way, as the purchases were reduced to levels compatible with the reduced sales. A serious problem of reduced sales not shown here but which will be dealt with later is that the ownership costs of the idle equipment still have to be met from the reduced contribution. The effects of reduced sales revenue come through in month 4.

Month 4 (with Trade Credit for Hirers and from Suppliers)

	Cash out (£)	Cash in (£)
Sales (2 items of equipment on hire)		nil
Workshop labour	750	
Purchases:		
Spares for 2 items of equipment for 1 month	nil	

	Cash out (£)	Cash in (£)
Consumables for 2 items of equipment for 1 month	nil	
Revenue from previous month's sales		2 000
Payments for previous month's purchases:		
Spares	400	
Consumables	200	
Totals	1 350	2 000

Net cash flow for month 4   + £ 650
Contribution to company for month 4   + £ 650
Number of idle items of equipment   3

Thus, the smaller sales revenue from month 3 has reduced the net cash inflow from £2 750 to £650.

If in month 5 the sales improve to 4 items of equipment on hire, the improved cash flows again take a further month to come through, as months 5 and 6 illustrate.

Month 5 (with Trade Credit for Hirers and from Suppliers)

	Cash out (£)	Cash in (£)
Sales (4 items of equipment on hire)		nil
Workshop labour	750	
Purchases:		
Spares for 4 items of equipment for 1 month	nil	
Consumables for 4 items of equipment for 1 month	nil	
Revenue from previous month's sales		2 000
Payments for previous month's purchases:		
Spares	400	
Consumables	200	
Totals	1 350	2 000

Net cash flow for month 5   + £ 650
Contribution to company for month 5   + £2 050
(Contribution = sales − workshop labour − spare − consumables)
Number of idle items of equipment   1

Month 6 (with Trade Credit for Hirers and from Suppliers)

	Cash out (£)	Cash in (£)
Sales (4 items of equipment on hire)		nil
Workshop labour	750	
Purchases:		
Spares for 4 items of equipment for 1 month	nil	
Consumables for 4 items of equipment for 1 month	nil	
Revenue from previous month's sales		4 000
Payment for previous month's purchases:		
Spares	800	
Consumables	400	
Totals	1 950	4 000

Net cash flow for month 6   + £2 050
Contribution to company for month 6   + £2 050
(Contribution = sales − workshop labour − spares − consumables)
Number of idle items of equipment   1

The cash flow in month 6 reflects the increase in sales experienced in month 5.

The cash flow and contribution for the first 6 months of this venture are therefore:

Month	1	2	3	4	5	6
Net cash flows	−£ 750	+£2 750	+£2 750	+£650	+£ 650	+£2 050
Contribution	+£2 750	+£2 750	+£ 650	+£650	+£2 050	+£2 050

It is seen from this that the contribution more closely reflects the fluctuating sales, whereas the effects of trade credit produce different cash flows.

So far the company has been purchasing just enough spares and consumables for each month's operations, but if it wished to build up stocks of spares and consumables, then the cash flows would be reduced in order to fund the stocks. Months 7 and 8 are examples of building up stocks to meet a future sales boom.

Month 7 (with Trade Credit for Hirers and from Suppliers)

	Cash out (£)	Cash in (£)
Sales (4 items of equipment on hire)		nil
Workshop labour	750	
Purchases:		
Spares for 8 items of equipment for 1 month	nil	
Consumables for 8 items of equipment for 1 month	nil	
Revenue from previous month's sales		4 000
Payment for previous month's supplies:		
Spares	800	
Consumables	400	
Totals	1 950	4 000

Net cash flow for month 7	+ £2 050
Contribution to company for month 7	+ £2 050
Number of idle items of equipment	1
Stocks:	
Spares for 4 items of equipment for 1 month	£ 800
Consumables for 4 items of equipment for 1 month:	£ 400

In month 7 spares and consumables for 8 items of equipment were purchsed, although only 4 were required to support the sales in the month. The 4 purchased in excess are held in stocks, as shown. Contribution is calculated on the revenue and cost of the sales.

Month 8 (with Trade Credit for Hirers and from Suppliers)

	Cash out (£)	Cash in (£)
Sales (4 items of equipment on hire)		nil
Workshop labour	750	
Purchases:		
Spares for 8 items of equipment for 1 month	nil	
Consumables for 8 items of equipment for 1 month	nil	
Revenue from previous month's sales		4 000

Payment for previous month's
supplies:

Spares	1 600	
Consumables	800	
Totals	3 150	4 000

Net cash flow for month 8	+ £ 850
Contribution to company for month 8	+ £2 050
Number of idle plant items	1

Stocks:

Spares for 8 items of equipment for 1 month:	£1 600
Consumables for 8 items of equipment for 1 month	£ 800

Thus, the build-up of stocks has reduced the cash flow to + £850.

To meet the expected sales boom, the company increases the total holding to 8 items of equipment. In order to service this, additional labour is required, which brings the workshop labour cost to £1 800.

Month 9 (with Trade Credit for Hirers and from Suppliers)

	Cash out (£)	Cash in (£)
Sales (8 items of equipment on hire)		nil
Workshop labour	1 800	
Purchases:		
Spares for 8 items of equipment for 1 month	nil	
Consumables for 8 items of equipment for 1 month	nil	
Revenue from previous month's sales		4 000
Payments for previous month's purchases:		
Spares	1 600	
Consumables	800	
Totals	4 200	4 000

Net cash flow for month 9	− £ 200
Contribution to company for month 9	+ £3 800
(Contribution = sales − workshop labour − spares − consumables)	
Number of idle items of equipment	nil

Stocks:
Spares for 8 items of equipment
  for 1 month   £1 600
Consumables for 8 items of
  equipment for 1 month   £ 800

The increase in sales has led to a negative cash flow, because the sales have led to increased trade credit which will not show as cash inflows until next month and the increased workshop costs had to be met immediately. Also, the level of purchases of spares and consumables was high in month 8 and cash must be found to fund the stocks. Month 10 cash flow will improve as the increased revenues come through. Such rapid expansion could, if continued, lead to overtrading in which the company's cash resources would not be sufficient to support the expansion.

Month 10 (with Trade Credit for Hirers and from Suppliers)

	Cash out (£)	Cash in (£)
Sales (8 items of equipment on hire)		nil
Workshop labour	1 800	
Purchases:		
Spares for 8 items of equipment		
for 1 month	nil	
Consumables for 8 items of		
equipment for 1 month	nil	
Revenue from previous month's sales		8 000
Payments for previous month's		
purchases:		
Spares	1 600	
Consumables	800	
Totals	4 200	8 000

Net cash flow for month 10   + £3 800
Contribution to company for month 10 + £3 800
Number of idle items of equipment   nil
Stocks:
Spares for 8 items of equipment
  for 1 month   £1 600
Consumables for 8 items of
  equipment for 1 month   £ 800

The cash flow in month 10 now reflects the higher revenue of the sales in month 9. If in month 11 some of the stock held is used to maintain the equipment rather than buying new spares and consumables, the cash flow would further improve in month 12.

Month 11 (with Trade Credit for Hirers and from Suppliers)

	Cash out (£)	Cash in (£)
Sales (8 items of equipment on hire)		nil
Workshop labour	1 800	
Purchases:		
Spares for 4 items of equipment		
for 1 month	nil	
Consumables for 4 items of		
equipment for 1 month	nil	
Revenue from previous month's sales		8 000
Payments for previous month's		
purchases:		
Spares	1 600	
Consumables	800	
Totals	4 200	8 000

Net cash flow for month 11	+ £3 800
Contribution to company for month 11	+ £3 800
Number of idle items of equipment	nil
Stocks:	
Spares for 4 items of equipment	
for 1 month	£ 800
Consumables for 4 items of	
equipment for 1 month	£ 400

Although 8 items of equipment were on hire, spares and consumables were only purchased for 4. The remaining required spares and consumables were taken from stock and the value of the stock was reduced.

Month 12 (with Trade Credit for Hirers and from Suppliers)

	Cash out (£)	Cash in (£)
Sales (4 items of equipment on hire)		nil
Workshop labour	1 800	
Purchases:		
Spares for 4 items of equipment		
for 1 month	nil	
Consumables for 4 items of		
equipment for 1 month	nil	

Revenue from previous month's sales		8 000
Payments for previous month's purchases:		
Spares	800	
Consumables	400	
Totals	3 000	8 000

Net cash flow for month 12	+ £5 000
Contribution to company for month 12	+ £1 000
Number of idle items of equipment	4
Stocks:	
Spares for 4 items of equipment for 1 month	£ 800
Consumables for 4 items of equipment for 1 month	£ 400

The cash flow increases to £5 000 in month 12 when the revenues are £8 000 due to the high sales in month 11 and payments for purchases are only £1 200 because half the spares and consumables used in month 11 were drawn from stock rather than purchased. This is known as de-stocking. The cash flows and contributions are summarised in Table 13.1.

Summary of the difference between cash and contribution in Table 13.1:

- The cumulative contribution for the 12 months is £27 400.
- The cumulative cash flow for the 12 months is £23 400.
- The difference between these is due to the trade credit and stocks.

Trade credit
Creditors:
    Outstanding credit received from suppliers:

Spares	£ 800
Consumables	£ 400
	£1 200

Debtors:
    Outstanding credit given to hirers:

4 items of equipment on hire	£4 000
Difference between debtors and creditors	£2 800

Table 13.1 Summary of cash flows, contributions and stocks for 12 months

Month	1	2	3	4	5	6	7	8	9	10	11	12
Contribution (£)	2 750	2 750	650	650	2 050	2 050	2 050	2 050	3 800	3 800	3 800	1 000
Net cash flow (£)	– 750	2 750	2 750	650	650	2 050	2 050	850	– 200	3 800	3 800	5 000
Stocks (£)	–	–	–	–	–	–	1 200	2 400	2 400	2 400	1 200	1 200
Reasons for cash flow fluctuations	Trade credit given to hirers and received from suppliers		Falling sales in month 3		Increased sales in month 5		Increase in stocks in month 7	Increase in sales leading to increasing trade credit and increase in workshop labour costs to support new sales and stocking in month 8			De-stocking in month 11 while sustaining high sales levels in month 11	

Stocks

Spares	£ 800
Consumables	£ 400
Total	£1 200

Thus, the summary of trade credit and stocks is:

Trade credit	£2 800
Stocks	£1 200
Total	£4 000

Of the £27 400 of contribution earned, the company has: £23 400 in cash (less overheads and ownership costs, etc.); £2 800 in debtors less creditors; and £1 200 in stocks.

The cash flows calculated in Table 13.1 were calculated on the basis of trading operations and require further adjustment before they represent the company cash flow. These other adjustments are head office or overhead expenses, the acquiring and disposal of capital assets, value added tax, corporation tax and dividends.

13.2.2   Overheads

Adjusting the cash flows and profits for overheads requires a projection of the head office overheads. Some of these expenses, such as salaries, will be monthly; others, such as telephone, electricity and rentals, may be monthly or quarterly; while business rates may be half-yearly. Table 13.2 shows the original contribution and cash flows from Table 13.1 with the overhead adjustment included.

The overhead projections in Table 13.2 are as shown. These have been stated in the month in which they occur, and the same figure in any month has been deducted from the contribution as well as the cash flow. The difference between the cumulative contribution of £27 400 and the cumulative cash of £23 400 is due to stocks and the difference between debtors and creditors, as previously explained. The effect of overheads is simply to reduce the contribution and cash available.

13.2.3   Value Added Tax

A company collects value added tax on receipt of payment from customers. The company also pays value added tax on the payment of suppliers for

Table 13.2 Summary of profit and cash flows for 12 months with overheads included

Month	Contribution before overheads deducted (Table 13.1) (£)	Cash flow before overheads deducted (Table 13.1) (£)	Overheads (£)	Contribution less overheads (£)	Cash flow less overheads (£)
1	2 750	− 750	150	2 600	− 900
2	2 750	2 750	10	2 740	2 740
3	650	2 750	40	610	2 710
4	650	650	10	640	640
5	2 050	650	10	2 040	640
6	2 050	2 050	190	1 860	1 860
7	2 050	2 050	10	2 040	2 040
8	2 050	850	10	2 040	840
9	3 800	− 200	40	3 760	− 240
10	3 800	3 800	10	3 790	3 790
11	3 800	3 800	10	3 790	3 790
12	1 000	5 000	40	960	4 960
Cumulative values	27 400	23 400	530	26 870	22 870

goods received. The difference is calculated and either paid to or received from the Customs and Excise office quarterly. Depending on the difference in value between the amount of goods bought and sold in a quarter, the VAT can result in an outflow or an inflow. If the VAT is owed by the company, the collection of VAT has been acting as a source of short-term funds. If a VAT refund is due to the company, then the company has been funding the difference, and the effect of VAT on the company's cash flow should be calculated.

13.3   CASH FLOW FORECASTS FOR TRADING OPERATIONS IN EQUIPMENT HIRE

13.3.1   Trade Credit

Equipment hire companies give and receive credit. The credit given to customers is typically 1 month, so that any equipment hired this month will have an invoice issued at the time of supply and payment will be expected in, say, 30 days. In preparing a cash flow forecast from the sales forecast, the cash in, or revenue, can be derived by inserting the appropriate time shifts between the month in which the sales occur and the month in which the cash is received. In deriving revenue, it is usual to assume that not all the customers will pay on the due date and that some will default. An analysis of the timing of receipts from customers will indicate which assumed time shifts are appropriate. Representative figures would show that 70% of customers pay within 1 month, 25% within 2 months and 5% within 3 months. It may be that payment is never received on some invoices, and this should be taken into account in the cash forecast if this proportion is significant. Thus, a sales forecast and derived cash in or revenue would be as illustrated in Table 13.3.

Similarly, credit would be received from suppliers of spare parts and consumables, and from the forecast of required spare parts, repair materials and consumables the appropriate time shifts would be used to determine the cash out. Labour and operatives would be paid weekly and the staff monthly.

Table 13.3 Sales forecast and derived cash revenue

Month	1	2	3	4	5	6	7
Sales forecast (£)	10 000	15 000	18 000	12 000	etc.	etc.	etc.
1 month delay (70%) (£)		7 000	10 500	12 600	8 400	etc.	etc.
2 month delay (25%) (£)			2 500	3 750	4 500	3 000	etc.
3 month delay (5%) (£)				500	750	900	600
Cash in (£)		7 000	13 000	16 850	13 650	3 900	600

Stocks and Work in Progress

As an equipment hire operation is not involved in manufacture, there is no substantial purchase of raw materials leading to stocks, nor is there any substantial work in progress. Stocks are mainly confined to spare parts and repair materials.

Sales Fluctuations and Overheads

As with any type of business, the equipment hire industry is subject to sales fluctuations. These, like the earlier examples, lead to increases in trade credit and can lead to overtrading in a sustained expansion. Stocking and destocking of spare parts and consumables takes place within the cycle of sales fluctuations. Figure 13.1 illustrates the variations in demand, as measured by new orders for the UK construction industry from 1980 to

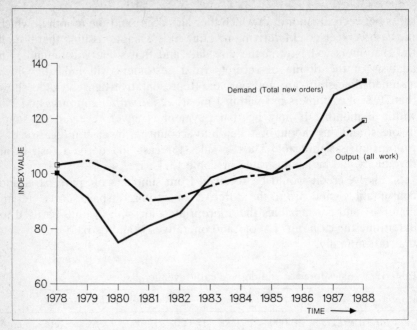

Figure 13.1 Demand and output: UK construction 1978–88 (source: DoE *Housing and Construction Statistics*, HMSO, 1989)

1988. Superimposed on this graph of demand is the overall output for the industry.

The buoyancy of the equipment hire industry is correlated with the output of the whole construction industry. Major changes in the level of output during the 1970s led to problems of overtrading on the rising markets to 1973, and later when lower levels of construction activity occurred, companies with disproportionate overheads were common. 1981 saw the low point in construction output, with sustained expansion occurring until the end of the 1980s. The general contractor is faced with cutting overheads as the market shrinks and the company's turnover declines, but the equipment hire company has another, more difficult, problem. This problem is that the equipment hire company has capital assets in the form of construction equipment which, if they become underutilised or hired out at an uneconomic rate, or both, quickly cease to give a return on capital. In such a situation companies would try to dispose of their equipment, but in a market slump this may not be possible. This is the basic risk in all equipment hire operations.

Forecasts

A short-term cash flow forecast is derived from the sales forecast and the aggregate cash requirements of all the various heads of account, divided into purchasing supplies, goods, labour, staff and overheads. The heads of account are the same as those used in the budget — namely workshop, transport and the various service departments of purchasing, administration, costing, hiring and finance. A tabular cash flow forecast may be constructed as illustrated in Table 13.4.

13.4   CASH FLOWS FOR THE PURCHASING AND ACQUISITION OF CAPITAL ASSETS

One feature of equipment hire companies that distinguishes them from general contractors is the high proportion of funds locked up in their capital assets. As much as 40–50% of the sales revenue may be dissipated on the ownership costs of the company's assets. The cash flow for the normal month to month trading as in Tables 13.2 and 13.4 should be combined with the company's capital expenditure budget, which includes disposals as well as purchases, to produce the cash flow statement.

As the capital expenditure budget is likely to produce the larger cash flows and will dominate the company cash flow, it is not unusual for the capital expenditure budget to be prepared first.

The method of acquisition determines the cash flows for individual items and Chapter 6 reviewed the various methods of acquisition: outright purchase, credit purchase, hire purchase or lease. All have different cash flow implications, and the cash flows of these various methods of acquisition can be summarised as follows:

Method of acquisition	Cash flow
Outright purchase	Single large down payment
Credit purchase	Deposit plus regular payments
Hire purchase	Deposit plus regular hire charges
Lease	No deposit but regular lease payments

Thus, leasing is the least difficult to provide for, because it should be paid from revenue. Hire purchase or some credit arrangement is the next easiest, since a down payment is required and the remaining charges can be met from revenue. However, the down payment may be borrowed and the loan repayments met from revenue. Similarly, outright purchase requires a large single payment of cash, but all of this or part of it can be borrowed and payments can be met from revenue.

Providing a cash flow statement that includes the effects of acquisition and purchasing requires a prediction of the proceeds received from disposals together with a forecast of the expenditure on acquisition divided into down payments, hire purchase or loan repayments, interest charges

Table 13.4 A tabular form for the construction of a cash flow forecast for trading operations

Month	1	2	3	4	5
Sales forecast					
1 month delay (70%) 2 month delay (25%) 3 month delay (5%)					
Revenue or cash in					
Workshop					

Goods purchased
  cash out
Labour payroll
  cash out
Staff payroll
  cash out
Overheads (rent, rates, phone, etc.)
  cash out					

Total cash out for workshop					
Transport department					

Goods purchased
  cash out
Labour payroll
  cash out
Staff payroll
  cash out
Overheads
  cash out					

Total cash out for transport department					
A cash-out flow for each section or main head of account in budget would be prepared					
Total cash out from all sections					
Total cash in from sales					
Net cash flow					
Cumulative cash flow					

and lease payments. The following example has been chosen so that it may be included with the cash flow calculated in Table 13.2.

EXAMPLE 13.2

Of the 5 items of equipment initially acquired by the company, 3 were leased and 2 were bought on hire purchase. The outright purchase price of the plant items is £5 000. The lease payments are £270 each per month for 24 months. The HP deposit on 1 plant item is £1 000 and the monthly payments are £385 for 12 months. In month 8, when the plant holding was increased to 8 units, this was done by selling one of the plant items bought by hire purchase for £4 000, paying the HP company £1 320, representing the outstanding HP payments, less the interest that would be charged for the remaining period of 4 months. The fleet was increased by a furher 4 leased items at a cost of £270 per month each. The cash flows for these acquisitions and disposals are shown in Table 13.5. Cash flow for acquisition and disposal is now added to the cash flows for normal trading from Table 13.2 and is shown in Table 13.6.

VAT has previously been described. Interest charges will be due if the negative cash flow is funded from loans or an overdraft, and corporation tax will be due after all costs and interest have been deducted from sales revenue. The time lags associated with corporation tax will vary from about 9 months to perhaps 24 months, and the payments due would be for previous trading years. Dividends would be paid out on the decision of the board of directors and are determined by the company's cash position, among other factors.

To complete the equipment hire company's cash flow forecast the analysis that was suggested in Table 13.4 needs to be extended as in Table 13.7.

Aspects that perhaps need further explanation are the recovery of the capital monies invested in equipment through hire charges and the implications of corporation tax.

13.4.1   Sales Revenue, Depreciation and Corporation Tax

In a hire company the sales revenue will be exclusively or predominantly made up of hire charges for hired equipment. These hire charges will have been set at the current going market rate but will have been chosen not only for market reasons, but also for economic ones. The 'economic' hire rate will include four elements: direct costs, indirect costs or overheads, ownership costs and profit. Chapter 8 dealt with such hire rate calculations and described the various methods of including the ownership costs. All of

Table 13.5 Cash flow for acquisition and disposal of equipment

Month	1	2	3	4	5	6	7	8	9	10	11	12
Equipment acquisition. 5 items. value (£)	25 000											
3 items by lease (£)	15 000											
2 items by HP (£)	10 000											
Lease payments (£)	810	810	810	810	810	810	810	810	810	810	810	810
HP deposit (£)	2 000											
HP payments (£)	770	770	770	770	770	770	770	770				
Equipment disposal (£)								4 000				
HP settlement (£)								1 320				
Equipment acquisition. 4 items. value (£)								20 000				
4 items by lease (£)								20 000				
Lease payments (£)									1 080	1 080	1 080	1 080
Total cash out (£)	3 580	1 580	1 580	1 580	1 580	1 580	1 580	3 980	2 275	2 275	2 275	2 275
Total cash in (£)								4 000				
Net cash flow (£)	−3 580	−1 580	−1 580	−1 580	−1 580	−1 580	−1 580	+20	−2 275	−2 275	−2 275	−2 275

Table 13.6   Cash flow for normal trading from Table 13.2, plus cash flows associated with plant acquisition and disposal from Table 13.5

Month	Net cash flow from Table 13.2 (normal trading) (£)	Net cash flow from Table 13.5 (acquisition and disposal) (£)	Sum or total net cash flow (£)	Cumulative cash flow (£)	Contribution, less overheads from Table 13.2, less net cash flow from acquisition and disposal (£)
1	− 900	− 3 580	− 4 480	− 4 480	− 980
2	2 740	− 1 580	1 160	− 3 320	1 160
3	2 710	− 1 580	1 130	− 2 190	− 970
4	640	− 1 580	− 940	− 3 130	− 940
5	640	− 1 580	− 940	− 4 070	460
6	1 860	− 1 580	280	− 3 790	280
7	2 040	− 1 580	460	− 3 330	460
8	840	+ 20	860	− 2 470	2 060
9	− 240	− 2 275	− 2 515	− 4 985	1 485
10	3 790	− 2 275	1 515	− 3 470	1 515
11	3 790	− 2 275	1 515	− 1 955	1 515
12	4 960	− 2 275	2 685	730	− 1 315

Note: Adjustments to this cash flow that would render it a true company cash flow would be for VAT, interest, corporation tax and dividends.

these are based on a concept of depreciation. This allowance for depreciation added into the hire rate is intended to ensure that the hire charges are large enough to recover the invested capital and an appropriate return on that invested capital. Thus, the monthly sales revenue includes this depreciation allowance. In the cash flow calculations undertaken in this chapter no attempt has been made to link the depreciation allowance included in the hire charges with the monies to be paid in ownership costs such as HP, loan repayments or lease payments. Although in calculating hire rates

$$\text{sales revenue} = \text{direct costs} + \text{indirect costs} + \text{depreciation} + \text{profit}$$

it is not adequate to say

$$\text{profit} = \text{sales revenue} - \text{direct costs} - \text{indirect costs} - \text{depreciation}$$

and that the monies available to meet ownership costs are depreciation plus retained profits.

Table 13.7   Headings for the cash flow forecast for acquisitions and disposals

			Month			
Item	*1*	*2*	*3*	*4*	*5*	*etc.*
Plant disposals						
Outright purchases						
Down payments						
Hire purchase payments						
Loan repayment						
Lease repayment						
Interest						
Value added tax						
Corporation tax						
Dividends						

Profit is not calculated in this way for tax purposes, as capital allowances are used to offset the cost of acquiring capital assets (see Chapter 14). Furthermore, the partitioning of the sales revenue in this way unrealistically suggests that the part of the revenue described as 'depreciation' is unnecessarily limited to be used in meeting ownership costs.

The reason that the sales revenue was left unpartitioned is that the funds can be deployed in any way to suit the company's operations. If the need is to meet the funding of debtors due to increasing sales or stocks or to buy a new item of equipment the surplus of revenue over cost can be deployed in the most advantageous way to suit the company. The company's need for cash is the criterion which dictates the use to which that cash available is put. The original method of calculating the hire charge which made allowances for depreciation was only a method of arriving at a realistic hire charge and has no influence on how the income is used. Thus, the cash flow analysis recommended in this chapter does not partition the revenue under the heading used to calculate the original hire charges.

The implications of corporation tax also need to be understood in the context of the depreciation element of sales revenue. If a company buys a capital asset for £1 000, UK tax legislation allows this to be offset against tax at the rate of 25% written down and in that year tax is charged on the remaining profits. In the first year this amounts to £250, to be set against tax. In the subsequent year 25% of the remaining capital sum of £750 (i.e. £187.50) is set against tax, and so on. The company's internal depreciation included in the hire charge is unlikely to recover the capital in exactly the same way or at the same rate. It may be that the company recovers the capital over 3 years including one-third of the capital in each year. Thus, in the year of purchase, if the company's profits were £1 200 (i.e. sales revenue less costs), the taxable profit would be £950, less interest charges, which is £1 200, less the capital allowances. If, in the next year, the company's profits (sales revenue, less costs) were £1 300, the taxable profit would be £1 112.50, less interest charges. As explained in Chapter 6, the method of acquisition influences the capital allowances. The internal depreciation included by the company in their hire charges has no bearing on the calculation of taxable profit.

The corporation tax implications from the three methods of acquisition — outright purchase, hire purchase and leasing — can be summarised as follows:

Method of acquisition	*Implications for corporation tax*
Outright purchase	A capital allowance of 25% written down of the purchase price can be set against profits.
Hire purchase	A capital allowance of 25% written down of the purchase price can be set against profits. The interest element of the HP charge is deductible from revenue before tax.
Leasing	No capital allowance is available but lease payments are deductible from revenue before tax.
Interest charges	Interest charges on loans and overdrafts are deductible from revenue before tax.

Thus, the method of acquisition which may be chosen for cash flow reasons has a bearing on the corporation tax due, which, in turn, affects the company's cash flow. The benefit of capital allowances can be derived only if there are adequate profits against which the allowances can be set.

13.5   CASH FLOW MANAGEMENT

This chapter has dealt with explanations of how and why cash flows vary and the need to forecast cash requirements. There has been an accompanying implication that the cash flow can be managed, which is so within limits. The variables of sales fluctuations, the amount of credit given, level of stocking, when to dispose of capital assets, when to acquire capital assets and by what method they should be acquired are all subject to some managerial control. However, there are limits.

13.5.1 Sales Fluctuations

It is theoretically easy in an expanding market to control sales growth and, hence, the increase in debtors. However, increasing sales may be difficult to resist but are likely to be restricted if this means acquiring more capital assets. It is more difficult to improve sales in a declining market: cutting hire charges may be an option but this could lead to unprofitable trading. Thus, if a major slump in the construction industry occurs, some equipment hire companies will inevitably cease to trade. A slump in the construction industry is out of the control of equipment hire companies.

13.5.2 Trade Credit

The amount of trade credit is set by the general trading conditions, and if a company sets shorter credit periods than competitors, the company may lose customers. Thus, trade credit is not wholly within the company's control. However, vigorous credit control can ensure that invoices are not allowed to remain unpaid for long periods beyond the normal credit given. A credit control system is a very important feature of a company's cash flow management. The company's credit controllers have some control, in that credit can always be refused if the customer is deemed unworthy of the risk.

With respect to credit received from suppliers, it is unlikely that a company could extend the normal credit arrangements without losing discount and jeopardising confidence in the company. Confidence is important in obtaining credit, overdrafts and loans.

13.5.3 Stocking

The level of stocks of spare parts, repair materials and consumables is a matter for equipment companies to decide. The equation that is being balanced is the cost of holding such stock against the risk of equipment remaining idle while spare parts are sought.

13.5.4 Disposal of Capital Assets

The timing of disposals is within the control of company managers, but the capital raised by such disposals is controlled more by general market conditions than by company manager's needs.

13.5.5   Acquisition of Capital Assets

The timing and the method of acquisition are both within the control of company managers, subject to delivery delays, etc.

13.5.6   Risk

Another need for cash that has not been considered so far is access to cash to cover risk. It is possible to plan for routine maintenance and, drawing on experience, it is possible to plan for repairs within reason, but it is not difficult to imagine situations where expensive repairs are required unexpectedly. The option of delaying repairs until cash becomes available is usually an undesirable situation, as this leaves the equipment item idle and ownership costs accrue whether the item is idle or not. Consequently, unexpected and expensive repairs often need to be undertaken immediately, and provision for such situations must be made. This implies the ready availability of cash for equipment not covered by an engineering insurance policy, and the ready availability of cash could suggest that cash may be idle and not working unless invested in short-term investments.

13.5.7   Cash, Deficits and Surplus

The cash flow illustrated in Table 13.6 showed that the monthly net cash flow in that example varied from − £4 480 to + £2 685. If this example were scaled up to a more realistic size, it would show that the company cash flow can swing from considerable deficits to substantial surplus. The company needs a source of funds to cover the deficits and a reasonable use for the surplus that will earn returns without leaving the cash idle. The traditional source of funds in excess of the company's own cash is overdrafts. The use of surplus cash includes short-term investments that can be quickly realised to meet cash requirements. In times of high inflation and high interest rates the contribution of such investments can be a source of considerable income to companies.

BIBLIOGRAPHY

Coombs, W. E. and Palmer, W. J. (1984). *The Handbook of Construction Accounting and Financial Management*, 3rd edn, McGraw-Hill

Drury, C. (1988). *Management and Cost Accounting*, 2nd edn, Van Nostrand Reinhold

Harris, F. C. and McCaffer, R. (1989). *Modern Construction Management*, 3rd edn, Blackwell Scientific

Hussey, R. (1989). *Cost and Management Accounting*, Macmillan

Lee, T. A. (1984). *Cash Flow Accounting*, Van Nostrand Reinhold

Mott, C. H. (1981). *Accounting and Financial Management for Construction*, Wiley

Padget, P. (1983). *Accounting in the Construction Industry*, Institute of Cost and Management Accountants

Samuels, J. M. and Wilkes, F. M. (1986). *Management of Company Finance*, 4th edn, Van Nostrand Reinhold

Sizer, J. (1989). *An Insight into Management Accounting*, 3rd edn, Penguin

Smith, J. E. (1981). *Cash Flow Management*, Woodhead Faulkner

Smith, J., Keith, R. and Stephens, W. (1988). *Managerial Accounting*, McGraw-Hill

Chapter 14
Financial Management

14.1 INTRODUCTION

An equipment hire or rental concern, like any other, usually has limited liability status, with either private or publicly quoted shares, depending upon its ability to secure a quotation on the Stock Exchange. A limited company, irrespective of its size, must file its annual financial accounts with the Registrar of Companies. To ensure that these give a true and fair view of the company's trading position, they are subjected to an annual audit by an independent firm of professional accountants appointed by the Department of Trade and Industry. If the company, on the other hand, is a subsidiary or department of a parent company, its trading position need only be included generally in the controlling company's accounts.

Beyond these legal requirements, however, the financial accounts provide a basis for measuring the profit made by the company and its overall financial performance during the year past. In addition, the accounts give shareholders information on the investment policy, borrowing and other details of interest to investors and creditors. In the annual report the main items presented are the Profit and Loss Account, the Balance Sheet and notes to the accounts.

14.2 THE PROFIT AND LOSS ACCOUNT

The Profit and Loss Account is a statement of the company's total profit or loss resulting from trading during the year. Its main features are the revenues generated from sales together with the costs incurred in producing the sales. The difference between the two values represents the profit or loss. After deduction of interest charges on borrowed capital and corporation tax, the resulting surplus is used to provide a dividend to shareholders and/or for reinvestment in the company as retained earnings.

EXAMPLE 14.1: A PROFIT AND LOSS ACCOUNT FOR 1990 (INCLUDING CAPITAL ALLOWANCES)

	£000s
Turnover (revenue)	
Hire of equipment and services	16 500
Costs	
Materials, labour, expenses (including depreciation), overheads	14 500
Net profit	2 000
Add depreciation back in (depreciation value agreed with tax inspector)	(2 000)
Assessable profit	4 000
Deduct capital allowances on plant and equipment (agreed with tax inspector)	2 500
Add profit from sales of fixed assets	(500)
Profit before taxation and interest	2 000
Deduct interest charges on borrowings (15% p.a.)	1 500
Profit before tax	500
Deduct corporation tax (say 35%)	175
Distributable profit after tax	325
Proposed dividend	225
Profits retained in business, C/F to Balance Sheet	100

Turnover

The company's turnover is the total value of sales of goods and services during the year, before costs are deducted. For a hire company or division this sum represents the total payments invoiced to clients for equipment hired or rented.

Assessable Profit for Corporation Tax with Capital Allowances

The assessable profit is obtained by subtracting the cost of sales from the turnover value. At present published accounts are not required to reveal details of the operating costs, which comprise the following elements: materials used — i.e. opening stock, plus purchases, minus closing stocks; wages, salaries and fees; expenses, excluding depreciation; administration overheads.

Depreciation and Capital Allowances

The principles of depreciation are discussed in Chapter 8. Internal depreciation rates are usually ignored for preparation of the Profit and Loss Account, as it is likely that capital allowances (i.e. writing down allowances), if available under tax legislation, would be more favourable. Capital allowances such as depreciation are deducted from the trading profit and the resulting sum is the net profit before interest charges on borrowed capital and tax. 100% capital allowances (i.e. the full cost of an item of new or used equipment set against profits during the year of purchase), for many firms, has had the net effect of virtually eliminating the need to pay corporation tax.

One effect of capital allowances can be merely to defer payment of taxes rather than avoid them, especially if there is no corresponding upsurge in profitability as a result of the purchase. Such an effect is best illustrated by a simple example of two identical companies having the same gross profit of £450, one favouring the use of capital allowances, and the other not. In both cases they buy an item of plant worth £1 000, which is written off in equal amounts over 4 years.

Company 1: Not Using Capital Allowances

	Year 1	Year 2	Year 3	Year 4
Profit before depreciation	450	450	450	450
Depreciation	250	250	250	250
Net profit	200	200	200	200
Tax at 35%	70	70	70	70
Profit after tax	130	130	130	130

Therefore total tax paid by Company 1 = £280

Company 2: Using Capital Allowances

	Year 1	Year 2	Year 3	Year 4
Net Profit	200	200	200	200
Depreciation	250	250	250	250
Assessable Profit	450	450	450	450
Capital Allowance	(1 000)	(550)*	(100)*	–
Taxable Profit (loss)	(550)	(100)	350	450
Tax at 35%	–	–	122.5	157.5

Therefore total tax paid by Company 2 = £280
* Capital allowance carried forward.

In both instances the total tax paid is £280, but with Company 2 this has been deferred.

In practice it is apparent that many hire companies try to avoid paying this deferred tax by continuing to buy new equipment: year 4 in the second example shows what could occur if such continual buying ceased. In this case the company would be liable to payment of taxes in excess of the net profit earned for the year. Therefore, not only do capital allowances provide a carrot, but also they provide a stick. A company could, in certain circumstances, be effectively trapped in the system by a slow, almost undetected, build-up of tax liabilities.

Interest Charges

Loan capital, other than shareholder's equity, usually incurs an annual interest charge. In effect, this represents an indirect cost on the business and is deducted as such from net profit for the calculation of corporation tax payment.

Profit before Taxation

The income from all the company's activities, including subsidiary and associated companies, is added to its own profit before tax. The total is called the *pre-tax profit*, upon which corporation tax is levied. The amount of tax due varies, depending on government policy and also on the size of company turnover. Currently, depending on the size of company, about 35% of profits are payable in corporation tax. After provisions have been made for payment of advanced corporation tax, the remaining tax liability can be deferred until the following year after declaration of the company report, and therefore provides a cheap source of short-term funds to aid cash flows.

Distribution of Profits after Tax

After the total distributable profits are determined, amounts may be apportioned to the various claimants. The order of priorities is set out in the articles of association of the company, with preference shareholders having first claim, the remainder of the distributed profits going to the ordinary shareholders. The profits awarded in this way are usually called the *dividend*. The dividend is recommended by the directors of the firm at the annual general meeting, who then vote to accept it or otherwise. The amount of dividend will naturally depend on the level of profits and may not be declared when profits are too low. Whenever healthy profits are available for distribution, the directors usually aim for an acceptable rate of return on the par (nominal) value of the share price.

Once the dividend has been paid, the remaining profit is retained in the business and used to finance further investment in plant and other assets.

14.3   THE BALANCE SHEET

The function of the balance sheet is to portray the financial position of the company on a specific date — for example, 31 December 1990. Because time is needed to prepare the Balance Sheet information, the details could change just before or immediately after publication — e.g. settlement of an account by a debtor. The balance sheet is a 'photograph' of a particular financial position, whereas the profit and loss account shows the record of achievement throughout the year.

The Balance Sheet is usually presented in tabular format separating capital employed from employment of capital. Alternatively, some companies prefer liabilities and assets to be clearly separated on different pages. In either case the two sets of measures must balance exactly. Thus, for example, any increase in cash must be counterbalanced by a decrease in some other asset or, alternatively, by an increase in liabilities. This principle applies throughout.

EXAMPLE 14.2: A BALANCE SHEET (WITH CAPITAL ALLOWANCES)

Using information given in the previous example together with the following company details, the Balance Sheet at 31 December 1990 may be prepared as follows.

Note 1: Information in Preparation of the Balance Sheet as at 31 December 1990

	£000
Value of fixed assets (see Note 2)	20 000
Cash at bank	910
Stock of materials and spares	1 000
Debtors	4 000
Bank overdraft	10
Creditors	3 000
Taxation (from P. & L. Account)	175
Proposed dividend	225

Profit and loss account transferred to general reserve   100
Interest paid on loans (15% p.a.)   1 500
Issue Share Capital   1 900
Reserves at 31 December 1989   10 000
Loan Stock (at 15% interest) at 31 December 1989   12 000
Loan Stock repaid during 1990   2 000
Capital allowances for plant agreed with Tax Inspector   2 500

Note 2: Valuation of Assets (Buildings and Equipment)

	£000	
Valuation at 1 January 1989	21 000	
Additions in year	10 000	
Disposals during year	(3 000)	
Valuation at 31 December 1989	28 000	28 000

Buildings and equipment depreciated
  according to company policy agreed with
  the Inspector of Taxes (capital allowances
  must be determined separately)

Depreciation

Cumulative up to 1 January 1989	6 000	
Provided for in 1989 (agreed with Tax Inspector)	2 000	
	8 000	
Net book value at 31 December 1989	20 000	20 000

Note 3: Capital Allowances

The book value of the firm's assets shown on the Balance Sheet is normally calculated by taking a realistic rate of depreciation. Because in this example capital allowances and not depreciation is used to calculate profit for corporation tax assessment, the difference between capital allowances and depreciation, like unrealised profit, should be allocated to the reserves. Furthermore, some accountants would show parts of these net allowances as deferred taxation, as it should be recalled that they will be liquidated when the asset is sold, or gradually eliminated as the asset is fully depreciated over its life.

In the example given, capital allowances on equipment purchases for 1990 are £2 500, whereas depreciation has been assessed at only £2 000. The net £500 has therefore been allocated to reserves.

Tabulated Balance Sheet as at 31 December 1990

	Depreciation			
	£000	£000	£000	
Employment of capital				
Fixed assets	28 000	8 000	20 000	20 000
Current assets				
Cash at bank			910	
Stock of materials			1 000	
Debtors			4 000	
			5 910	5 910
Current liabilities				
Bank overdraft			10	
Creditors			3 000	
Taxation due			175	
Proposed dividend			225	
			3 410	3 410
Net current assets				
(working capital)			2 500	
			22 500	22 500
Capital employed				
Issued Share Capital			1 900	
Reserves				
(1) P. & L. Account		100		
(2) Net capital allowances		500		
(3) As at 31.12.89		10 000		
		10 600	10 600	
Loan stock			10 000	
			22 500	22 500

Note: Because the interest paid on loans has been transacted it is not shown as a liability.

14.4 LOAN CAPITAL

Short-term loans are usually classified with current liabilities, but medium- and long-term loans are placed with the capital employed. Loans may take several forms, of which debentures and loan stock form the most important sources. Such lenders of capital are creditors of the company, and not owners as are shareholders.

14.4.1   Long-term/Medium-term Finance

Long-term finance is that capital required for 5–10 years, either to start the business or to carry out expansion programmes. Broadly, the capital is used to purchase buildings, plant and equipment and to carry stocks of materials. The risks to the lender are high because of the time scale involved and, consequently, only established firms are generally considered by the lending institutions. Some of the more important sources of long-term capital are shown in Figure 14.1.

Loans are not easy to obtain. Lenders of such capital often request the borrower to provide a proportion of the finance from internal sources and, in addition, require convincing evidence that the loan capital can be secured against an asset with profitable expectations. The rate of interest, period of loan and capital repayments vary according to the lenders' request.

Debentures are offered for sale by the company, and are requests for loan capital with a fixed annual interest payment and life. Clearly, the company must show that it is able to pay the interest, and the capital itself will usually be mortgaged against the company's assets. Loan stock and debentures rank ahead of shareholders in entitlement for payment.

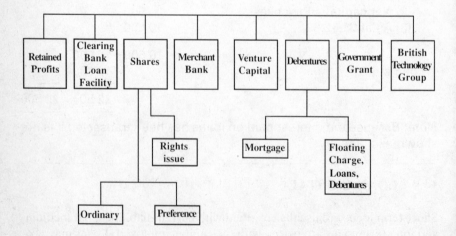

Figure 14.1   Sources of long-term finance

14.4.2   Short-term Finance

The firm, when established, often needs short-term capital to overcome immediate cash flow problems. Materials have to be purchased, equipment hired, labour and subcontractors paid, and so on, before payment is received for the finished product or service. Furthermore, capital may be required to smooth out the strains on cash flow resulting from rapid fluctuations in the market demand for the company's goods. Many sources of short-term finance are available to ease the situation, but naturally the firm must be well managed and profitable before the lending institutions will consider any loan application. The main sources are shown in Figure 14.2, but the clearing bank overdraft facility is the most important source. However, a leasing facility from a finance house is also an important source of either long- or short-term funding of plant acquisitions. The method is described in Chapter 6 and operates more like a rental payment than a loan.

Sources of most types of capital, their advantages and disadvantages and the costs involved as summarised in Table 14.1. Table 14.2 illustrates capital acquisition methods used by typical construction companies.

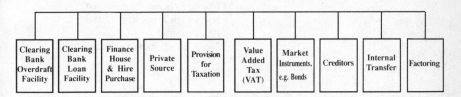

Figure 14.2   Sources of short-term finance

14.5   WORKING CAPITAL

Working capital is represented by the difference between current assets and current liabilities and is locked up in a continuous cycle, as shown in Figure 14.3.

EXAMPLE 14.3: WORKING CAPITAL REQUIREMENTS

A construction company owns and operates equipment, grouped into a separate division responsible for generating its own turnover in the market place. The turnover of this division is £20 million pounds per year, broken down as 45% plant ownership costs, 15%

Table 14.1 Summary of capital sources

Source	Finance	Advantages
Bank	Overdraft	(1) Usually cheapest source (2) Quickly arranged (3) Flexible (4) No minimum (5) Renewable (6) Interest paid only on usage (7) Sometimes available unsecured
Bank or finance house	Short-term loan	(1) Term commitment by loan institution (2) Competition between lending houses, especially for hire purchase (3) Relatively quickly arranged (4) Can be used in conjunction with overdraft facility (5) Sometimes available unsecured
Finance house	Hire purchase facility	(1) Inexpensive and specially arranged (2) Payments fixed over term agreed (3) Ideal for short-term requirements (4) Normally overdraft facility not affected (5) Capital allowances against corporation tax available immediately (6) Not classed as borrowings
Finance house	Lease facility	(1) Similar advantages to hire purchase (2) Overgeared firm can acquire resources without affecting Balance Sheet
Clearing banks Merchant banks	Medium term loan	(1) Term commitment by lender (2) Capital and interest repayment can be arranged to suit borrower's future cash flow position (3) Size of loan may be small or large, especially from clearing banks (4) Inflation reduces real cost over time (5) Fixed interest charge can sometimes be negotiated

Disadvantages	Costs
(1) Subject to changes in government economic policy (2) Repayable on demand (3) Subject to changes in bank policy (4) Tempting to use for funding long-term purchases	(1) Floating interest charge at base rate plus 1–4% (2) May incur commitment fee
(1) Generally more expensive than overdraft (2) Term commitment and funds may therefore be idle if forecast for funds inaccurate (3) Tends to require security against other assets	(1) Floating interest charge at base rate plus 2–5%
(1) Expensive (2) Subject to government economic policy changes, but never retrospectively (3) Defaults usually rigorously prosecuted (4) Interest rate quoted may be misleading, because of period of payments and compound interest calculations (5) Purchased assets not legally passed over to lender until final payment	(1) Interest fixed at time of negotiations at finance house base rate plus 4–5%
(1) Ownership does not pass to lessee and therefore capital allowances not available (2) Values not reflected in balance sheet assets, and so might give distorted impression of firm's capabilities	(1) Similar to hire purchase costs, but user must be aware of possible commitments to pay for mandatory maintenance and repairs (2) Tax concessions foregone must be compared with HP alternative
(1) Usually higher interest charge than shorter-term finance (2) Long-term commitment which may require short-term borrowings to finance interest payment if cash flows became distorted from forecasts (3) Negotiation fee likely (4) Legal costs also often incurred	(1) Fixed or variable interest charge set at $1\frac{1}{2}$–4% above 6 months inter-bank rate

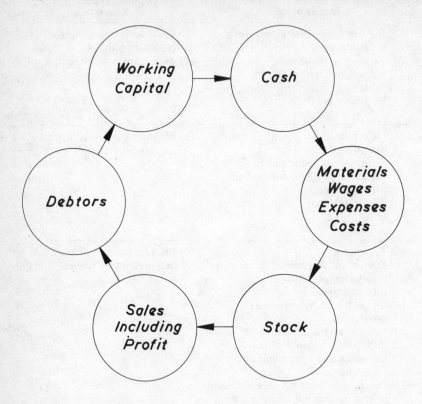

Figure 14.3 The working capital cycle

Table 14.2 Capital acquisition methods
used by typical construction companies

Method	Total (%)
Hire purchase	30
Lease	20
Outright purchase	50
	100

for materials and 10% for wages for maintenance and transport, 20% for overheads incurred in maintaining the depot establishment and administration facilities; and 10% profit. On average, the company keeps 3 months' material spares in stock and is allowed 3 months credit by suppliers: wages are paid weekly. Hirers of the company's equipment (i.e. debtors) are usually allowed up to 2 months to pay. Overheads must be met monthly. The ownership costs comprise the capital repayment and interest charges for hire purchase agreements to be paid monthly. Determine the minimum working capital required.

Working Capital Calculations
Materials (15% × £20 millions = £3 millions p.a.)

	Time factor
Materials held in stock	3 months
Credit to customers	2 months
	5 months
Less credits from suppliers	3 months
	2 months

$$\therefore \text{Materials} = \frac{2}{12} \times 3\ 000\ 000 = £500\ 000$$

Labour (10% × £20 millions = £2 million p.a.)

	Time factor
Credit to customers	2 months
Less 1 week arrears on wages	0.25 months
	1.75 months

$$\therefore \text{Labour} = 2\ 000\ 000 \times \frac{1.75}{12} = £292\ 000 \text{ approx.}$$

Overheads (20% × £20 millions = £4 millions p.a.)

	Time factor
Credit to customers	2 months
Less 1 month arrears on salaries, rents, rates, etc.	1 month
	1 month

$$\therefore \text{Overheads} = 4\,000\,000 \times \frac{1}{12} = £330\,000 \text{ approx.}$$

Ownership costs (45% × £20 millions = £9 millions p.a.)

	Time factor
Credit to customers	2 months
Less time delay on hire purchase payment	1 month
	1 month

$$\therefore \text{Ownership costs} = 9\,000\,000 \times \frac{1}{12} = £750\,000$$

Working Capital Requirements

The above figures can now be added together to calculate the total working capital requirements:

Materials	£ 500 000
Labour	£ 292 000
Overheads	£ 330 000
Hire purchase	£ 750 000
	£1 872 000

Such capital needs would usually be financed by means of an overdraft negotiated with a bank or, alternatively, with private funds. However, it can be seen that if equipment had been purchased with internal funds rather than by hire purchase, the working capital needs would be £0.75 millions less. The equivalent sum would then remain in the business and offset the working capital needs.

14.6   ASPECTS OF THE FINANCIAL ACCOUNTS

14.6.1   Ratio Analysis

The main short-term techniques for managerial control are cash flow forecasting, budgetary control and costing, described in Chapters 12 and 13. Unfortunately, however, they provide absolute figures which are of marginal value in monitoring the long-term profitability and short-term liquidity of the company. Regular internal ratio analysis using the financial accounts offers a complementary post mortem to these standard procedures.

Working Capital Ratios

$$(1) \text{ Current ratio} = \frac{\text{Current assets}}{\text{Current liabilities}}$$

$$(2) \text{ Acid test} = \frac{\text{Cash and debtors}}{\text{Current liabilities}}$$

Most accountants look for ratios of 2:1 and 1:1, respectively. Using the Balance Sheet given earlier, it can be seen that:

$$CR = \frac{5\ 910\ 000}{3\ 410\ 000} = 1.73$$

$$AT = \frac{910\ 000 + 4\ 000\ 000}{3\ 410\ 000} = 1.44$$

In this hypothetical company the current ratio (CR) is a little low and the acid test (AT) a little high. This situation could be improved by increasing stocks by £$\frac{1}{2}$ million and reducing creditors and debtors, respectively, by £1 and £1$\frac{1}{2}$ million, to give the following values:

$$CR = \frac{4.91}{2.41} = 2.03$$

$$AT = \frac{3.41}{2.41} = 1.41$$

Profitability and Operating Ratios

$$(1) \ \frac{\text{Net profit before tax}}{\text{Turnover}} = \frac{2\ 000\ 000}{17\ 000\ 000} = 11.8\%$$

$$(2) \ \frac{\text{Net profit before tax}}{\text{Capital employed}} = \frac{2\ 000\ 000}{22\ 500\ 000} = 8.9\% \text{ (primary ratio)}$$

$$(3) \ \frac{\text{Turnover}}{\text{Capital employed}} = \frac{17\ 000\ 000}{22\ 500\ 000} = 0.76\% \text{ (turnover ratio)}$$

The turnover ratio can provide telling information about the capital structure of the company. For example, the plant hire company is more similar to manufacturing with its turnover ratio approaching 1, whereas with most construction companies the ratio is more often nearer to 6, the difference being accounted for by the huge capital investment required in plant. The ratio could, of course, be increased by leasing rather than owning equipment.

Other ratios

$$\frac{\text{Debtors}}{\text{Turnover}} \times 12 = \frac{4\ 000\ 000}{17\ 000\ 000} \times 12 = 2.82 \text{ months}$$

This means that customers are given nearly three months to pay.

$$\frac{\text{Creditors}}{\text{Purchases}} \times 12 = \frac{3\ 000\ 000}{12\ 500\ 000} \times 12 = 2.88 \text{ months}$$

This means that suppliers give nearly 3 months credit.

The ratios suggest that the company is applying a sensible policy with respect to both suppliers and customers, and a deeper investigation may also yield important details regarding productivity, using the following ratios:

(1) $\dfrac{\text{Turnover}}{\text{Number of employees}}$

(2) $\dfrac{\text{Profit}}{\text{Number of employees}}$

(3) $\dfrac{\text{Plant and equipment value}}{\text{Number of employees}}$

(4) $\dfrac{\text{Profit}}{\text{Plant and equipment value}}$

14.7   CAPITAL GEARING

Capital gearing is defined as the ratio of fixed return capital (FRC) to ordinary share capital, FRC being preference shares, debentures and loan stock. A company with a ratio exceeding 1 is described as a highly geared company.

When a company can expect confidently to make a constant level of high profits over a number of years, then it is wise to raise some of the capital by means of debentures or loan stock and so improve the yield to the ordinary shareholder. Such a technique is known as leverage, and the effects are demonstrated in the following example.

EXAMPLE 14.4

The capital structures for two plant hire companies are given in Table 14.3 and the profits available for interest payments and dividends for a range of company performance levels are given in Table 14.4.

(1) Calculate and comment on the capital gearing ratios for each of the companies.
(2) Calculate the dividend available to the ordinary shareholders and comment on the earnings per share at each level of profit. Corporation tax is 50%.

Table 14.3 Capital structures

	Company (£1000)	Company (£1000)
Ordinary shares (£1)	310	70
7% Preference shares	90	90
10% Loan stock (debentures)	–	240

Table 14.4 Performance levels

	1	2	3	4	5	6
Profit (£)	10 000	12 000	18 000	20 000	25 000	30 000

Solution Part (1)

	Company X	Company Y
10% Debentures	–	240
7% Preference shares	90	90
FRC	90	330
Ordinary shares	310	70
Capital gearing ratio	$\dfrac{90}{310} \times 100 = 29\%$	$\dfrac{330}{70} \times 100 = 470\%$
	Low-geared	High-geared

Company A is low-geared and may be able to declare a dividend to ordinary shareholders, even at low levels of profit. Company B is high-geared and must therefore ensure that profits are stable and regular and, above all, sufficient to meet the interest due on debenture holdings. However, the highly geared company should be able to yield a higher rate of return to ordinary shareholders when profits are high.

Solution: Part (2)

Table 14.5 shows that Company X is able to pay the full dividend to preference shareholders at all times: even the ordinary shareholder received payment at all the given levels of profit.

Table 14.6 shows that Company Y is in a much more difficult position while profits are poor. In fact, for the first profit period a loan or bank overdraft must be negotiated in order to meet the interest charges on the debenture stock. The full dividend to preference shareholders cannot be met until profit level 5, when there is also a small amount available for distribution to ordinary shareholders. However, the high profits at levels 5 and 6 yield a far better return to the ordinary shareholders than in the low-geared company and this trend will continue for all increased levels of profit.

Table 14.5 Company X (low-geared)

| | Performance level (£) | | | | | |
	1	2	3	4	5	6
Ordinary shares (£1)	310	310	310	310	310	310
Preference shares	90	90	90	90	90	90
Debentures	–	–	–	–	–	–
Total capital employed (£000)	400	400	400	400	400	400
Profit before tax	10	12	18	20	25	30
Corporation tax at 35%	3.50	4.20	6.30	7.00	8.75	10.50
Profit available as dividend	6.50	7.80	11.70	13.00	16.25	19.50
Dividend on preference shares	6.30	6.30	6.30	6.30	6.30	6.30
Profit available to ordinary shareholders	0.20	1.50	5.40	6.70	9.95	13.20
Profit per ordinary share	0.06p	0.5p	1.74p	2.16p	3.2p	4.2p
Return on ordinary share capital	0.06%	0.5%	1.74%	2.16%	3.2%	4.2%
Return on total share capital	1.63%	1.95%	2.9%	3.25%	4.0%	4.8%

Table 14.6 Company Y (high-geared)

	Performance level (£)					
	1	2	3	4	5	6
Ordinary shares (£1)	70	70	70	70	70	70
7% preference shares	90	90	90	90	90	90
5% debentures	240	240	240	240	240	240
Total capital employed (£000)	400	400	400	400	400	400
Profit before tax	10	12	18	20	25	30
Interest on 10% debentures	12	12	12	12	12	12
Profit available	− 2	nil	6	8	13	18
Corporation tax at 35%	nil	nil	2.1	2.8	4.55	6.3
Profit available as dividend	nil	nil	3.9	5.2	8.45	11.7
Dividend to preference shareholders	nil	nil	3.9	5.2	6.30	6.3
Profit available to ordinary shareholders	nil	nil	nil	nil	2.15	5.4
Profit per ordinary share	nil	nil	nil	nil	3p	7.7p
Return on ordinary share capital	nil	nil	nil	nil	3%	7.7%
Return on total share capital	nil	nil	2.4%	3.2%	5.28%	7.3%

If the preference shares were issued as cumulative, then in the case of the highly geared situation additional dividend would be paid out in the good years to make up for the lost years, but at the expense of the ordinary shareholder.

14.8 PLANT PROFITABILITY

It has been argued that it is more advantageous for a plant department to be operated as a profit centre and not on the service principle, as is often the case in a construction company. This is best illustrated by the following example, using financial data for a plant-holding construction company.

EXAMPLE 14.5

A construction company has an annual turnover of £20 millions, of which one-fifth is from equipment operated by its plant department. The expected profit is 20% of the capital employed and the company is typical of the construction division in that capital is turned over eight times per year, whereas for a plant department the turnover ratio is commonly unity or thereabouts.

	£ millions		
	Parent company	Construction division	Plant division
Turnover	20	16	4
Capital employed	6	2	4
20% profit on capital employed	1.2	0.4	0.8
Profit expressed as a percentage on turnover	$\dfrac{1.2}{20} \times 100 = 6\%$	$\dfrac{0.4}{16} \times 100 = 2.5\%$	$\dfrac{0.8}{4} \times 100 = 20\%$

If the plant division, for instance, had made only £0.5 millions profit, then, for the company as a whole, the construction division would have to make £0.7 millions profit, or 4.375% on turnover, to achieve 6% overall on turnover. Thus, any shortfall in plant profitability requires a monumental effort of the construction division to redress the balance.

14.9   ACCOUNTING FOR INFLATION

During the past 10 years or so, the annual level of inflation has sometimes exceeded 10%. Consequently, results portrayed by the historical method of accounting described earlier in this chapter are not 'true and fair', as required by the Companies Act. The main effects of inflation are as follows:

(1) Assets are undervalued. If, for example, a company purchases an excavator for £10 000, then, 10 years later, the same machine may cost £20 000 because of inflation. The valuation shown in the accounts, however, will be based on £10 000. The company's assets and the shareholders interest are, therefore, undervalued. Also, more importantly, because the depreciation is based on the purchase date figure, profits will be overstated. When capital allowances (i.e. 100% depreciation during the year of purchase) are in operation, this latter detrimental effect of inflation is significantly lessened.
(2) Stocks of materials are undervalued. The change in the value of stocks, between opening and closing dates in the accounts, may have been caused by both a volume difference and a change in prices. This latter

aspect is not taken into account in calculating the cost of sales and so, in times of rising prices, the profit is overstated and extra tax is paid.

(3) Companies which finance their operations by loan capital will gain over companies that are self-financing because monetary debts will depreciate in real value. For example, in real terms £1 000 borrowed and repayable ten years later, is equivalent to £385 if the rate of inflation is 10% p.a.

As a result of high inflation, it became obvious that the historical system of accounting was becoming unsatisfactory and disadvantageous to the commercial sector of the economy. Several bodies subsequently put forward proposals for dealing with the problem and in 1980 the Accounting Standards Committee published guidance notes for preparing financial accounts with inflation. However, these are not mandatory under the Companies Acts, and taxation and other government requirements are still largely based on the historical accounts, with some exemptions allowed for stock appreciation. The Accounting Standards Committee has recommended that the traditional historical accounts should also be presented with the inflation-adjusted ones: in this way the effects on profits of a combination of inflation and the present requirements for payment of corporation tax can be more clearly monitored. The necessary adjustments to the historical accounts are largely those described under points (1), (2) and (3) above, and the reader is referred to the Committee's guidance notes for further advice.

BIBLIOGRAPHY

Accounting Standards Committee (1986). *Accounting for the Effects of Changing Prices: A Handbook*

Briston, R. J. (1981). *Introduction to Accountancy and Finance*, Macmillan

Glautier, M. W. E. and Underdown, B. (1986). *Accounting Theory and Practice*, Pitman

Glautier, M. W. E., Underdown, B. and Clarke, A. C. (1985). *Basic Financial Accounting*, Pitman

Hindle, T. (1985). *Pocket Banker*, Blackwell

Nedved, S. C. (1973). *Builders' Accounting*, Newnes-Butterworth

Pizzey, A. (1988). *Accounting and Finance*, Cassell

Chapter 15
Using Computers in Managing Construction Equipment

15.1 INTRODUCTION

The management of an equipment division within a company, an equipment company within a group of companies, or an independent equipment hire company encompasses all the general management tasks faced by most companies. As in other types of companies, the use of computers has become established in basic bookkeeping or accounting functions such as payroll and ledgers. Some equipment companies have extended the use of computers into management accounting and management information systems. Examples of these uses include the maintenance of an asset register, which is particularly valuable because of the high value of capital assets represented in a large number of equipment items. Other examples include equipment location reports, cost and revenue reports, and variance analyses, which compare income with budgeted income and actual usage with planned usage. Stock control and maintenance record keeping are also examples of management information applications.

Other uses of computers in equipment management include financial appraisals where financial modelling systems are used to explore the cash flow and profitability prospects of various proposals.

The larger companies have had access to computers for many years, and the applications developed by these companies have naturally developed to suit the larger computers. More recently, the availability of small microcomputers has extended the use of computers to smaller companies and many of the applications referred to above are available on these smaller machines.

This chapter briefly reviews the computer applications that are particularly relevant to the management of construction equipment.

15.2   COMPUTER SYSTEMS

Until the late 1970s most computing work of the type undertaken by equipment companies employed a large computer system known as a main frame computer. Since then the so-called micro revolution has taken place and many of the applications that were hitherto available only on large computers are available on microcomputers. Many companies still use and will go on using large computers because it suits their operations and management organisation, but the arrival of the microcomputer has extended the number of users of computers and no company is now excluded from access to a computer simply on grounds of capital cost.

15.3   COMPUTER APPLICATIONS IN THE MANAGEMENT OF EQUIPMENT

Computers are used in the management of equipment in the following applications:

(1) Basic accounting or book-keeping: (a) payroll; (b) purchase ledger; (c) sales ledger; and (d) nominal ledger.
(2) Management accounting and information: (a) asset register; (b) purchase and disposal analysis; (c) weekly hire charges reports; (d) period hire charges reports; (e) plant locations; (f) revenue and cost reports; (g) plant utilisation reports; (h) stock control; and (i) maintenance records.
(3) Financial appraisals.

15.3.1   Survey: The Use of Computers by Equipment Hire Companies

In 1987 in an attempt to determine the extent of computer use by equipment companies, 120 companies were contacted, divided evenly between the eight regions defined by the Construction Plant Hire Association. Comprehensive responses to a detailed questionnaire were returned from almost 30% of the firms. (See Harrington, 1987; Harrington and Harris, 1988).

Response

(1) About 50% of respondents claimed to be using computers, with installations growing at roughly 5% per year. The stand-alone micro-computer had almost 50% share of the market.
(2) Invoice generation and payroll administration and financial statements continued to be popular. Applications also included updating of hire rates, hire location, stock and costs recording, and providing asset register and maintenance information.
(3) Almost 75% of computerised respondents had obtained 'bought-in' computer software rather than designed 'in house'. However, about half these had only purchased standard packages such as word process-ing and spreadsheets, specifically Lotus 123 and Supercalc. Purpose-designed packages seemed to be limited to database applications, particularly in updating stock records matched to fleet usage.
(4) Spending intentions on computers and software were buoyant but mostly small-scale and generally less than £5 000 for the coming year. Nevertheless, such sums represented considerable capital investment for many of the small companies in the hire sector. However, one of the larger groups was intending to outlay £100 000 on an entirely new system. Overall, it appeared that plant departments and companies were currently in a vigorous phase of computerising their activities. Only very few relatively small firms appeared not to be interested.

15.3.2 Basic Accounting or Book-keeping

Payroll

Among the labour-intensive industries, construction is perhaps the one with the most complicated payroll requirements. Very few firms in construction pay all their employees on the same basis: instead there is a mixture of remuneration by piecework, and hourly, daily or weekly wages, as well as monthly salaries. As a result, the computerisation of payroll always produces considerable benefits.

There is a great number of general-purpose payroll packages available for practically every computer in the market. As with all commercial software, payroll programs can be bought or leased for implementation on the user's own computer.

An important aspect of payroll programs is that they deal with some-thing which is not stable. Taxation changes every year, often more than once; so does national insurance. Pension and bonus schemes also change, and so does the form and the kind of information that the Inland Revenue, and other authorities, want presented to them at the end of each year. A company's employees often work on different contracts within a single period, and the distribution of their time among these contracts changes every day. Payroll packages have to cope with this fluidity.

The basic facilities of a payroll system are:

- Payroll creation, the creation and amendment of fixed details.
- Payroll calculation, calculating the weekly and monthly pay of employees.
- Printing payslips, P35, P11 and non-employees list.
- Amend tax and N.I. rates and pensions.
- Payment by cash, cheque or credit transfer.

Payroll systems based on timesheet information allow labour costs to be allocated to each job, thus providing information for job costing.

Sales, Purchase, Nominal Ledger and VAT

The facilities offered in sales and purchase ledger systems are: (1) accounts creation, amendments and deletions; (2) accounts postings; (3) cash postings; (4) period statements and remittance advices. Derived from these basic accounting functions, the reports that are available include accounts over credit limit, accounts overdue, accounts turnovers and customer lists. Linked to the sales and purchase ledgers are VAT maintenance systems. These collect and analyse VAT collected and paid, enabling the total amount due to be calculated. The nominal ledger systems maintain a set of nominal accounts.

An important aspect of having the basic accounting functions computerised is that the information contained in the accounting system can be made available to other functions such as equipment and job costing and stock control. The information contained in the payroll system and in the sales and purchase ledgers is relevant to these other functions and in many accounting systems offers the important facility of feeding information to these other functions. Two important chains can be identified: purchase ledger–payroll–job costing and purchase ledger–order processing–stock control.

15.3.3 Management Accounting and Information

The applications of computers in management accounting and information are less standardised than the basic accounting functions. Consequently, there are fewer computer packages available for purchase. Also, it is a group of applications that more closely reflects the management and organisation of a company. Consequently, this group of applications, more than the basic accounting functions, have been developed within companies for their own use. Nevertheless, several software houses do offer systems purposely designed and suitable for equipment companies. The applications that come in this group are reviewed as follows.

Asset Register

Assets accounting deals with assets which, usually, have a long lifespan, extending over several accounting periods. Computer programs for assets accounting provide, at any one time, (1) an assets register with straightforward information on what assets the contractor has, and where they are; (2) their written-down value, and depreciation to date; and (3) their replacement cost.

Information provided on the replacement cost is perhaps the most valuable, as it forms the essential test on the wisdom, or otherwise, of the depreciation policy being adopted. Most asset register systems offer a range of depreciation methods, including straight line and declining balance.

Thus, reports from asset register systems include:

- A list of plant items grouped by type of equipment.
- Inventory numbers.
- Value.
- Depreciation to date by usual methods.
- Depreciation for this year.

Separate reports will include technical details of each item of equipment held.

The importance of an asset register for the equipment company is that it maintains detailed information relating to the items of equipment held in a form that supports the production of the company accounts and in a form that allows managers to consider their depreciation and disposal policies. The depreciation policy is particularly important to an equipment company, because the depreciation element, or capital cost, forms as much as 50% of a hire rate. The purchase and disposal of equipment is equally important to the continued well-being of a company.

Purchase and Disposal Analysis

Linked to the asset register is a purchase and disposal analysis, which is a maintained record of equipment items bought and sold. This aids the preparation of annual accounts and assists companies in managing the size of their equipment fleet.

Equipment Reports

The equipment hire company requires its own set of management reports to inform the company managers and to allow them to take the various day-to-day decisions.

The management reporting systems dealing with equipment include:

- Weekly hire charges.
- Period hire charges.
- Equipment locations.
- Revenue and cost reports.
- Equipment utilisation reports.

Here the computer systems are being employed in handling relatively large quantities of data regularly, to provide the equipment manager with the information he requires.

The two key reports are the Equipment Utilisation Report and the Equipment Cost and Revenue Report. These two reports provide essential information for the running of any large holding of equipment. Essentially they are designed for the equipment hire company to monitor their costs and revenues and, hence, their profits. But they are equally essential for the construction company holding equipment in order that they may manage the equipment subsidiary as a profit centre whose main revenues come from internal charges.

The Equipment Utilisation Report

This information is usually produced monthly and is intended to be a statement of the equipment's actual usage in comparison with the budgeted (or planned) usage. Using the principles of management accounting, the equipment utilisation report tends to present the information in the form of variances. These variances compare the actual performance with the anticipated performance, the performance being measured in terms of the number of hours on hire and the 'price' (or revenue) obtained for these hours of hire.

Equipment Cost and Revenue Report

This report, also produced at least monthly, is the comparison of costs and revenues. The costs listed in such a report would refer to direct costs associated with the equipment item, and so the net revenue remaining still has to pay for company overheads before being properly classified as 'profit'.

The other reports that are available include:

Weekly and Period Hire Charges

This reporting system records the current hire charges for all equipment and the dates when hire charges were changed.

Equipment Locations

Equipment location reports record the current location of equipment on hire and the length of time at that location. The location of idle equipment is also monitored in reports of this type.

Stock Control

Stock control is described in more detail in Chapter 9. In equipment management, stock control relates mainly to the supply of spare parts and consumables. The purpose of keeping stock is to supply the fleet with the items it needs, but this need must be tempered with economy. Overstocking causes locked-up capital but understocking may cause a repair to be delayed and involve expensive down-time. Good stock control aims at achieving a balance between overstocking and understocking. Stock recording and control systems monitor stock issues and delivery notes, and perform stock taking and prompt reordering when minimum levels of stock of a particular item are reached.

Stock control systems are usually linked to purchasing systems and hold details of the various suppliers, the purchase order details for the suppliers and the invoices. Stock control systems offer savings by optimising the level of stock held and holding up-to-date information on the stocks and reducing the manpower committed to stock taking.

Maintenance Records

The maintenance records described in Chapter 9 can be held on computer file for ease of update and reference.

15.3.4   Other Applications: Financial Appraisals

A common application relating to equipment management that is not part of day-to-day company management routines is financial appraisals. A number of investment appraisal programs are available to perform such tasks.

These investment appraisal programs provide a range of calculations which enable a proposed investment to be evaluated. The investment in this case is a capital sum from which revenue, or some other form of return, is expected, and the evaluation determines whether these returns are enough to justify the capital investment. The adequacy of these investments is measured by profitability, not just simple profit. The most common profitability measures, based on discounting techniques, are net present value and yield or rate of return. Other measures produced by these programs, which do not rely on discounting, are the payback period and the average annual rate of return. All these methods of measuring profitability are described in Chapter 5.

The more sophisticated programs offer facilities which enable the modelling of the estimates and calculations that determine the investment projects' cash flow. Using this modelling facility, the estimates can be varied and the sensitivity of the investment project's profitability to certain estimates can be determined. An example of these financial appraisals would be as follows.

Figure 5.1 (page 61) shows a model of the relationship between the cash flows that go into calculating a rate of return. The rate of return is defined as the *discounted cash flow yield*. Given estimates, based on historical records, and experienced guesswork for each of the elements in the model, the expected DCF yield can be calculated. However, two of the key elements in this calculation are the hire rate and the utilisation factor. The 'market' controls both the hire rate and the utilisation factor and therefore neither can be readily estimated with accuracy. Using some of the existing software packages, it is relatively easy to calculate the rate of return for a range of utilisation factors and for a range of hire rates. By way of example, this has been done for an actual case, and the results are plotted and shown in Figure 5.3 (page 68) The interesting aspect of Figure 5.3 is that it tells you that the equipment needs to be utilised at 60% for hire rate 1 before it breaks even, and if the company specified its minimum rate of return as 10%, this particular item of equipment would need to be utilised at 82%. It is now a market judgement as to whether this item of equipment can be hired out for the required time to be profitable. The exciting aspect of this type of calculation is that after it has been done as a means of evaluating whether the equipment item should be in the fleet or not, the same data or graph can be used in a management sense to set targets for salesmen.

The other major use made of these evaluation programs arises in the decision whether to purchase outright, to use hire purchase or to lease. The capital payments of each of these three methods of acquisition are different and occur at different times. Also, the implications with respect to corporation tax and capital allowances are different.

The capital allowances set against tax for equipment are 25% written down. Outright purchase, therefore, attracts this allowance. However, to take advantage of these allowances, the company must be making profits. Thus, the method of acquisition is affected by the capital available and the company's current profits. These factors can be 'modelled' in systems, which are essentially programs that permit the manipulation of cash flows. Thus, the method of acquisition that best suits the company at the particular time can be evaluated.

Most of the calculations described in this section are capable of being undertaken manually with calculators. However, the use of computers serves to make the calculations more thorough, in that, the exercise for computer analysis having been set up, it is easy to explore it more thoroughly and to explore more alternatives.

15.4   AVAILABLE COMPUTER SYSTEMS

The Harrington (1987) survey attempted to determine the capabilities of different computer systems relevant to the hire sector and catalogued 11 systems available in 1987. The hardware catalogued ranged from £2 000 to £10 000, with software in many cases doubling the cost.

Software available to equipment hire firms is constantly being revised and new products frequently appear on the market. Thus, any list of software can quickly become outdated.

BIBLIOGRAPHY

Anon. (1986). The easy invoice. *Plant Managers Journal*, February
Anon. (1987). The screening of Britain. *Which Computer*, February
Anon. (1987). Whose hands on the reins? *Practical Computing*, July
Baldry, C. (1984). Computers for plantmen. *Plant Managers Journal*, May, June
Harrington, M. F. (1987). Computer Trends in the UK Plant Hire Industry, project report for the Master's degree, Loughborough University of Technology, Department of Civil Engineering
Harrington, M. F. and Harris, F. C. (1988). Trends in the plant hire industry. *Construction Computing*, Spring

Chapter 16
Overseas Operations

16.1 INTRODUCTION

Over the years, Western companies have increasingly undertaken construction work overseas, particularly in developing countries. Projects have included housing, roads, power stations, water supplies, water treatment facilities, mining, pipelines and other infrastructure activities. More recently contracts for electromechanical complexes, steelworks and other industrial developments have also been common. Most have required changes in management, construction and, especially, procurement practices in dealing with operations remote from the home base.

16.2 REMOTE REGIONS

A remote region usually implies locations in uninhabited territory and sparsely populated areas lacking basic infrastructure and utilities such as good transport networks and communications. However, unfamiliar conditions can occur almost next door, resulting from different languages, and political, economic or historical developments. Nevertheless, in general, operations will mainly involve changed geography, climate, and legal and/or unaccustomed social practices. The 'comfortable' home arrangements of immediate deliveries, spares 'off the shelf', ready advice and abundant experienced staff, etc., are unlikely to be available.

16.3 TYPES OF PROBLEM

Many of the problems to be faced in supplying and operating equipment in remote regions are common to well-developed societies, but usually require considerably more management effort. Typically, wrong decisions and choices are made and, as a consequence, already overstretched resources become exacerbated, the situation being further compounded by scheduling, ordering, despatch and, especially, delivery delays. Indeed, systems designed for a developed home market often break down when exposed to the vagaries of poor communications and infrastructure of a remote site. Other difficulties commonly arise through the need to raise or receive funds in a foreign currency not readily acceptable to the Foreign Exchange market. Also, import controls and customs regulations can be complicated and very restrictive, particularly regarding imported spare parts and especially where the host country claims to have a manufacturing capability.

Obtaining the right personnel to both manage and operate equipment usually proves difficult. Indigenous staff often require considerable training, both in the skills for the job itself and in communicating with the home base and systems of the company, including the need to converse in a foreign language. Expatriates often predominate at supervisory levels and, unless highly paid, thoroughly selected and prepared for an overseas environment, are unlikely to perform satisfactorily. Furthermore, having to cope with local customs and practices commonly borders on corruption and can become very distorted unless entrusted to very reliable staff.

More specifically, poor communication facilities, such as telephone systems, postal services and even radio transmission, lead to frustration and errors. Regarding the plant itself, inexperienced staff usually cause extra maintenance, increased breakdowns and failures, etc., further compounded by damages to equipment in transit, either through bad roads and transport or even lack of adequate inspection of imports at source. Also, extra quantities, especially of spare parts, may need to be ordered, to avoid stock shortages. Furthermore, the remoteness of overseas work frequently exposes inefficient manufacturers and suppliers of equipment, with many unprepared for dealing with difficult logistics and markets.

16.4   EQUIPMENT REQUIREMENTS

16.4.1   Support Items

The types of equipment needed to execute an overseas project will clearly depend upon the nature of the work. However, support facilities can be more specifically mentioned and would typically include:

(1) Concrete batching and mixing equipment.

(2) Transport vehicles, such as trucks, 4-wheel-drive Land Rovers, utility vans and wagons, etc.

(3) Welding and fabrication shops, equipped with welding and burning gear, workshop equipment, grinders, etc.

(4) Maintenance workshops, for plant and vehicles, together with spare parts stores.

(5) Power generators, including a standby water storage tank, fuel dumps, etc. A water treatment capability might also be necessary.

(6) Waste disposal facilities for sewage – e.g. septic tanks, incinerators. Indeed, on large long-term sites proper sewage treatment equipment may be worth installing.

(7) Furnished site offices, canteens, stores and compound, plant yard. Many sites will also need to provide furnished accommodation for some of the work force, which might have to be quite elaborate for expatriates, including recreational facilities.

16.4.2   Equipment in General

Wherever possible, the equipment acquired or designed for an overseas contract should be versatile, with the capacity to change one duty for another — e.g. tractor to carrier. Standardisation can assist here in facilitating switching and cannibalisation. Sensitive equipment generally needs protection from the heat or cold, dust, humidity, wind, etc., with instruments requiring special attention, particularly during transport. Temperature-sensitive materials should be kept under controlled conditions in a cold store or air-conditioned rooms.

Above all, regular maintenance is essential, even to the extent of turning on electric motors, etc., regularly to keep moving parts free. A plentiful supply of stocks of consumables, such as steel plate, structural steel pieces, pipework, hosing, fittings, etc., is clearly essential besides normal plant spares, and, above all, a well-protected and well-policed stores is essential. Evidence indicates that about 10–15% of plant value in spares in usually needed to be stocked on remote sites. However, the spares and stocks requirements may fluctuate markedly over a contract, and therefore a good-size initial supply is prudent until usage patterns have been ascertained. Local supplies may be available but may be inferior if not manufactured in one of the major industrial countries. This whole question impacts on maintenance, since, if the major manufacturers of equipment have no service backup in the country, either local fitters, electricians, mechanics and technicians have to be trained or imported expatriates have to be employed at considerable expense. In either case, decisions regarding provision of appropriate spares and workshops must be carefully thought out.

16.5　SUPPLY AND DELIVERY

While many of the above points affect the selection of equipment and the arrangements needed to effectively manage and operate, supply, packaging, transport, agents, customs, finance, insurance, and so on, are equally important.

16.5.1　Pre-shipment Inspection

Equipment ordered is commonly specified to certain standards — for example, British Standards, DIN, etc — and many developing countries insist on pre-shipment inspections charged to the supplier, to ensure that the conditions are met before items leave the source country. The procedures requested may involve simply counting of items and dimensional checks, or more thorough inspections, including testing of materials, components, complete items of equipment, etc., in accordance with the particular specification standard. Other aspects could include monitoring the export price against like items sold on the home market.

16.5.2　Transport to Site

Transport over long distances through unfamiliar territory and other countries en route clearly needs special attention and planning. Wherever transit periods exceed several months and different means of carriage, strong packaging and preferably containerisation is obviously necessary. The choice of transport method itself depends upon several factors — for example, roads, while generally economically competitive, usually involve long journey times and can be fraught with problems, such as poor road surfaces, lack of signposting, pilferage, banditry, and unreliable drivers and/or haulage firms. Also, when at journey's end, special temporary roads may need to be constructed, especially for the very remote construction site.

Alternatively, shipment over long sea routes, while generally cheap, is very slow but is commonly preferred for large plant items, to avoid dissembly demanded by container sizes. However, even with shipping, containerisation needs link-ups with trucks for ease of handling. A further consideration is the quality of facilities at the inloading port. Increasingly, rail transport is becoming competitive over long distances especially where national rail systems are well organised — for example Trans-Siberia, Trans Canada, and systems in parts of Africa and Europe.

Air freight is usually the quickest means of transport but also the most expensive. Heavy or bulky items are usually unsuitable, with large units often needing dissembling. Occasionally very specialist transport can be considered, such as helicopters, and even more exotic means such as snowmobiles, pack animals, and so on.

Finally, the courier should not be overlooked for small items, facilitating considerable shortening of customs delays.

16.5.3   Documentation and Bureaucracy

The bureaucracy of the importing country can be complex and certainly needs to be understood. For example, in addition to the specification and quality inspections mentioned above, other barriers such as tendering and trading procedures, legislation relating to labour employment and the environment, road transport regulations and restrictions regarding particular imports can cause problems. The advice of a local agent would certainly be helpful for the newcomer to the market, and every effort should be made to pre-empt difficulties. Much good advice can be obtained from British government and specific industry export organisations, such as the British Overseas Trade Board, export intelligence services and the Export Group for the Constructional Industries. As far as documentation is concerned, a local representative in the destination country is essential, not least to deal with language interpretation. Indeed, a local agent of this sort may be necessary in each country through which the goods have to pass. Good practice also usually requires employing a shipping agent to expedite and co-ordinate orders and deliveries, and to be responsible for packing lists, bills of lading, shipping, customs forms, currency payments, and so on. A good shipping agent will normally encourage marking of items with part numbers and the provision of detailed descriptions on the documentation to aid both checking and inspection, but also to enable follow-up enquiries when matters become aggravated through pilferage and/or transport deterioration.

16.6   FINANCE AND PAYMENT

Equipment purchased direct for overseas contracts usually requires the organising of import licences and, of course, the mechanisms of payment, as the host country is unlikely to have equipment available in a sophisticated hire sector similar to that in the UK, for instance. Furthermore, much plant may have to be fully depreciated during the contract and simply left in the host country, although in some countries small markets for the sale of second-hand equipment exists. Import priorities in developing countries are commonly solved by rationing through a waiting list, sometimes tempered by importance of use of the currency — e.g. equipment over consumer goods. Some transactions in local currency may be available via the foreign exchanges, but for very poor or ill-managed economies, few opportunities exist to exchange 'hard' (i.e. DM, dollars, £s, francs, etc.) for 'soft' currency, since such currency can really only be used to make purchases in the local economy. Some international bartering may be possible, and today organisations specialising in exchanging commodities for Western consumer and capital goods do exist.

Much difficulty can be avoided when projects are funded by aid agencies such as the World Bank, UN, ODD, etc.; here a soft loan is usually involved — i.e. a loan in a hard currency with low interest rate or easy payment terms, with the stipulation being sourcing of equipment from a particular country. Finally, import duties should not be overlooked — i.e. they vary considerably from country to country, even though international agreements may have been negotiated through GATT (General Agreement and Tariffs and Trade). Some countries waive the import duty if the equipment is to be re-exported at the end of its duties.

16.7   CONCLUSION

Clearly, when embarking on any overseas contracts, obtaining knowledge about the likely conditions to be faced is vital. Much advice is available in the home country through government export agencies, trade associations, R & D organisations, shipping agents, private consultancy firms, etc. However, reconnoitring the local environment is paramount and in addition, for many countries, local agents, distributers, contacts, etc., are absolutely essential: in some cases forming partnerships or joint companies may be justified in order to secure knowledge of (1) financial and banking procedures; (2) import legislation, quality specifications and standards; (3) customs, shipping, delivery and distribution procedure; (4) maintenance expertise and labour management.

BIBLIOGRAPHY AND ORGANISATIONS

British Overseas Trade Board, Victoria Street, London
Employment Conditions Abroad Ltd, Devonshire Street, London
Export Group for the Constructional Industries, King Street, St James, London
Exports Credits Guarantee Department, London
Moore, A. B. (1984). *Marketing Management in Construction*, Butterworths

Predicasts Information Services, Orpington, Kent
Urry, S. A. and Sherratt, A. F. C. (1980). *International Construction*, Construction Press

Appendix: Interest and Time Relationships and Tabulations for Interest Rates of 10% and 15%

A.1 INTEREST AND TIME RELATIONSHIPS

Cash flows are transfers of money. Positive cash flows are transfers into a project or scheme, and negative cash flows are transfers of money out of a project or scheme. If a company purchased an item of equipment, the purchase price and the runnings costs would be negative cash flows. The revenue from hire and the resale would be positive cash flows. Cash flows can be individual lump sums or a recurring series — that is, the same cash flow recurring each period. The manipulation of both types of cash flow, the lump sum and the recurring series, is achieved by the use of derived relationships between interest and time. The derived relationships are tabulated as factors which can be used to achieve these manipulations. The expressions used to calculate these factors are given later in the example tables for the interest rates 10% and 15%. The expressions are also given in the following six explanation examples, which are presented in the same order as that in which they are tabulated. The expressions are easily incorporated into computer programs of programmable calculations, and the use of tables is diminishing.

A.1.1 Compound Amounts

If a sum of money, £1 000, is invested for some years — say 8 — at an interest rate of 10%, the sum of money that can be withdrawn at the end of that time would be:

$$£1\ 000 \times 2.143\ 5 = £2\ 143.50$$

2.143 5 is the compound amount factor, taken from the tables or calculated from the expression $(1 + i)^n$, where i is the interest rate 0.1 (for 10%) and n is the number of years, 8. This process is represented in Figure A.1, to illustrate the reward received for money loaned or invested.

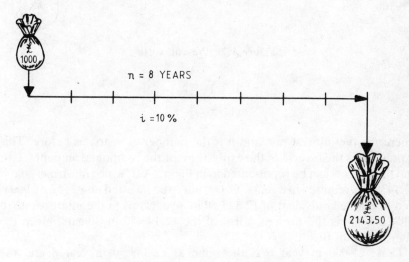

Figure A.1   Compound amount

A.1.2   Present Worth

The inverse of calculating compound amounts is calculating the present worth. If a sum of money, £2 143.50, is required in 8 years' time the capital sum that would be required to be invested today to generate this amount, given an interest rate of 10%, would be £1 000:

$$£2\ 143.50 \times 0.466\ 50 = £1\ 000$$

0.466 50 is the present worth factor, taken from the tables or calculated from the expression:

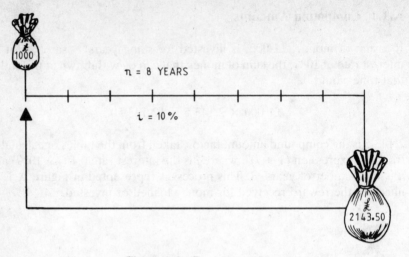

Figure A.2   Present worth

$$\frac{1}{(1 + i)^n}$$

where i is the interest rate and n is the number of years, as before. This expression is the inverse of the expression for the compound amount factor and the process can be represented as in Figure A.2, which illustrates that if £2 143.50 is required in 8 years, £1 000 has to be invested now. The £1 000 is said to be the equivalent of £2 143.50 in 8 years, given the interest rate of 10%. £1 000 is the present worth of the £2 143.50 in 8 years, given the interest rate of 10%.

Thus, £1 000 in year 0 is the same as £2 143.50 in year 8 and the difference of £1 143.50 is the interest earned in the intervening 8 years. This process of converting the £2 143.50 in year 8 to £1 000 in year 0 is known as discounting. This discounting process is very widely used, because it provides a convenient method of converting future cash flows to a common base date and provides a means of comparing cash flows of different magnitude occurring at different times.

It is important to remember that the present worth of a future cash flow is the capital sum that would need to be invested today to generate that future sum.

A.1.3   Compound Amount of a Regular or Uniform Series

If a sum of money, £100, is invested regularly for, say, 6 years, the sum available at the end of the 6 years — given an interest rate of 15% — is £875.30, given by:

$$£100 \times 8.753 = £875.30$$

The 8.753 is the uniform series compound amount factor, taken from the tables or calculated from the expression:

$$\frac{(1 + i)^n - 1}{i}$$

where i is the interest rate and n is the number of years.

Each sum is invested for a different period of time, and the compound amount could be calculated by using the compound amount factor for each of the six individual cash flows and summing the results. The uniform series compound amount factor allows this to be done in one step, as illustrated in Figure A.3.

Figure A.3 also shows that the cash flows are included in the calculation at the end of the year or period in which they occur. This is the assumption on which the expression was derived and is known as the 'end of period convention', which states that all cash flows take place at the end of the period.

A.1.4   Sinking Fund Deposit

The inverse of the compound amount of the uniform series is to find how much should be deposited each year or period in order to generate a certain sum at the end of the time. If £875.30 is required at the end of year 6 and it is intended to provide this sum by saving or investing a sum for each of the next 6 years at an interest rate of 15%, the sum to be saved would be £100:

$$£875.30 \times 0.114\ 23 = £100$$

0.114 23 is the sinking fund deposit factor, taken from tables or calculated from the expression:

$$\frac{i}{(1 + i)^n - 1}$$

where i is the interest rate and n is the number of years. This expression is the inverse of the uniform series compound amount factor. This factor is used in calculating the monies to be taken from revenue to repay borrowed capital or to replace plant.

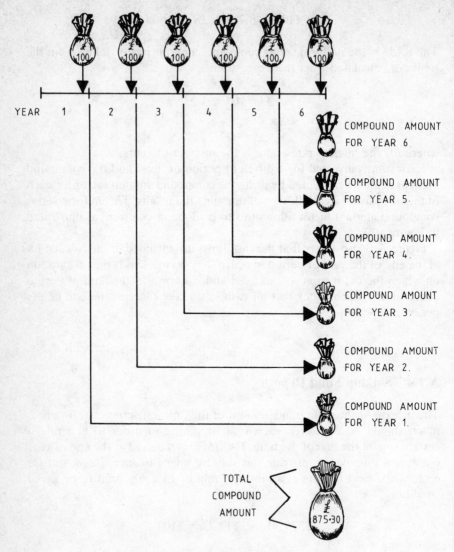

YEAR   1   2   3   4   5   6

COMPOUND AMOUNT
FOR YEAR 6.

COMPOUND AMOUNT
FOR YEAR 5.

COMPOUND AMOUNT
FOR YEAR 4.

COMPOUND AMOUNT
FOR YEAR 3.

COMPOUND AMOUNT
FOR YEAR 2.

COMPOUND AMOUNT
FOR YEAR 1.

TOTAL
COMPOUND
AMOUNT

£
875·30

Figure A.3   Compound amount of a uniform series

A.1.5   Present Worth of a Regular or Uniform Series

As the present worth factor calculated the present worth for a lump sum, this factor calculates the present worth of a series of cash flows recurring for a number of years or periods.

The present worth of £100 each year for the next 4 years, given an interest rate of 15%, is £285.49:

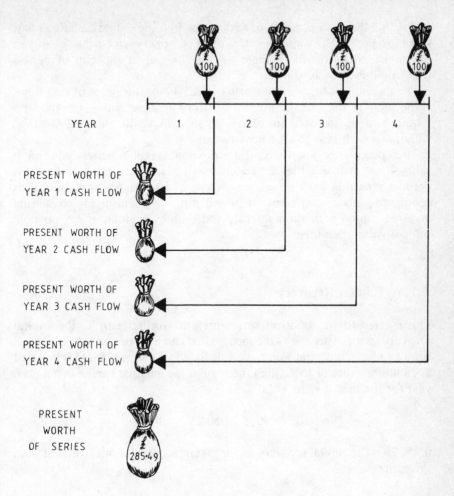

Figure A.4   Present worth of a uniform series

$$£100 \times 2.854\ 9 = £285.49$$

2.854 9 is the uniform series present worth factor, taken from tables or calculated from the expression:

$$\frac{(1 + i)^n - 1}{(1 + i)^n i}$$

where i is the interest rate and n is the number of years. It should be noted that the expression is the uniform series compound amount factor multiplied by the present worth factor. £285.49 could have been found by calculating the present worth of each of the four individual cash flows and summing them. The uniform series present worth factor achieves this in one step, as illustrated in Figure A.4. The use of the end of period convention is also illustrated.

The meaning of the present worth of £285.49 for this series of cash flows is the same as that for the present worth of a single sum — i.e. the sum required to be invested in year 0 at 15% that will allow £100 to be withdrawn each year for the next 4 years.

This process is widely used to discount regular series into lump sums — i.e. to capitalise the regular series. It is important because it permits recurring cash flows to be converted to capital sums and thus compared with capital sums. It provides the mechanism for comparing invested capital with running costs and thereby evaluating any possible trade-offs between them.

A.1.6   Capital Recovery

The inverse to the uniform series present worth factor is the capital recovery factor. This allows the income that can be taken on a regular basis from an invested capital sum to be calculated. If a sum £285.49 is invested at an interest rate of 15%, then the regular income that can be taken each year for the next 4 years is £100:

$$£285.49 \times 0.350\ 26 = £100$$

0.350 26 is the capital recovery factor taken from tables or calculated from the expression:

$$\frac{i(1 + i)^n}{(1 + i)^n - 1}$$

where i is the interest rate and n is the number of years. This expression is the inverse of the uniform series present worth factor.

The importance of this calculation is that it allows the conversion of capital sums into annual sums. This provides an alternative means of comparing capital with running costs. For example, if £285.49 is not invested but used to purchase an item of equipment, then the cost of that equipment item can be regarded as £285.49 or as an annual cost of £100 for 4 years, the annual cost being set equal to the amount of income that the investor is deprived of because the capital is locked up in the equipment. This calculation is useful in comparing hiring with purchasing or in calculating the 'capital' element in hire rates.

A.2   TABULATIONS OF INTEREST AND TIME RELATIONSHIPS FOR INTEREST RATES OF 10% AND 15%

Table A.1   Interest factors for 10%

Year or period (n)	$(1 + i)^n$ Compound amount of a single sum	$\dfrac{1}{(1 + i)^n}$ Present value of a single sum	$\dfrac{(1 + i)^n - 1}{i}$ Compound amount of a uniform series	$\dfrac{i}{(1 + i)^n - 1}$ Sinking fund deposit	$\dfrac{i(1 + i)^n - 1}{i(1 + i)^n}$ Present worth of a uniform series	$\dfrac{i(1 + i)^n}{(1 + i)^n - 1}$ Capital recovery
1	1·100 0	0·909 09	1·000	1·000 00	0·909 0	1·100 00
2	1·209 9	0·826 44	2·099	0·476 19	1·735 5	0·576 19
3	1·330 9	0·751 31	3·309	0·302 11	2·486 8	0·402 11
4	1·464 0	0·683 01	4·640	0·215 47	3·169 8	0·315 47
5	1·610 5	0·620 92	6·105	0·163 79	3·790 7	0·263 79
6	1·771 5	0·564 47	7·715	0·129 60	4·355 2	0·229 60
7	1·948 7	0·513 15	9·487	0·105 40	4·868 4	0·205 40
8	2·143 5	0·466 50	11·435	0·087 44	5·334 9	0·187 44
9	2·357 9	0·424 09	13·579	0·073 64	5·759 0	0·173 64
10	2·593 7	0·385 54	15·937	0·062 74	6·144 5	0·162 74
11	2·853 1	0·350 49	18·531	0·053 96	6·495 0	0·153 96
12	3·138 4	0·318 63	21·384	0·046 76	6·813 6	0·146 76
13	3·452 2	0·289 66	24·522	0·040 77	7·103 3	0·140 77
14	3·797 4	0·263 33	27·974	0·035 74	7·366 6	0·135 74
15	4·177 2	0·239 39	31·722	0·031 47	7·606 0	0·131 47
16	4·594 9	0·217 62	35·949	0·027 81	7·923 7	0·127 81
17	5·054 4	0·197 84	40·544	0·024 66	8·021 5	0·124 66
18	5·559 9	0·179 85	45·599	0·021 93	8·201 4	0·121 93
19	6·115 9	0·163 50	51·159	0·019 54	8·364 9	0·119 54
20	6·727 4	0·148 64	57·274	0·017 45	8·513 5	0·117 45
21	7·400 2	0·135 13	64·002	0·015 62	8·648 6	0·115 62
22	8·140 2	0·122 84	71·402	0·014 00	8·771 5	0·114 00
23	8·954 3	0·111 67	79·543	0·012 57	8·883 2	0·112 57
24	9·849 7	0·101 52	88·497	0·011 29	8·984 7	0·111 29
25	10·834 7	0·092 29	98·347	0·010 16	9·077 0	0·110 16
26	11·918 1	0·083 90	109·181	0·009 15	9·160 9	0·109 15
27	13·109 9	0·076 27	121·099	0·008 25	9·237 2	0·108 25
28	14·420 9	0·069 34	134·209	0·007 45	9·306 5	0·107 45
29	15·863 0	0·063 03	148·630	0·006 72	9·369 6	0·106 72
30	17·449 4	0·057 30	164·494	0·006 07	9·426 9	0·106 07
35	28·102 4	0·035 58	271·024	0·003 68	9·644 1	0·103 68
40	45·259 2	0·022 09	442·592	0·002 25	9·779 0	0·102 25
45	72·890 4	0·013 71	718·904	0·001 39	9·862 8	0·101 39
50	117·390 8	0·008 51	1 163·908	0·000 85	9·914 8	0·100 85

All examples presented in this appendix are based on each period being 1 year. A period can be any other unit of time – e.g. quarters, months or weeks. Care is required to ensure that the interest rate i and the unit of time for each period are compatible. For example, if the time period is months, then the interest rate must be a monthly rate.

Table A.2   Interest factors for 15%

Year or period (n)	$(1 + i)^n$ Compound amount of a single sum	$\dfrac{1}{(1 + i)^n}$ Present worth of a single sum	$\dfrac{(1 + i)^n - 1}{i}$ Compound amount of a uniform series	$\dfrac{i}{(1 + i)^n - 1}$ Sinking fund deposit	$\dfrac{(1 + i)^n - 1}{i(1 + i)^n}$ Present worth of a uniform series	$\dfrac{i(1 + i)^n}{(1 + i)^n - 1}$ Capital recovery
1	1·150 0	0·869 56	1·000	1·000 00	0·869 5	1·150 00
2	1·322 4	0·756 14	2·149	0·465 11	1·625 7	0·615 11
3	1·520 8	0·657 51	3·472	0·287 97	2·283 2	0·437 97
4	1·749 0	0·571 75	4·993	0·200 26	2·854 9	0·350 26
5	2·011 3	0·497 17	6·742	0·148 31	3·352 1	0·298 31
6	2·313 0	0·432 32	8·753	0·114 23	3·784 4	0·264 23
7	2·660 0	0·375 93	11·066	0·090 36	4·160 4	0·240 36
8	3·059 0	0·326 90	13·725	0·072 85	4·487 3	0·222 85
9	3·517 8	0·284 26	16·785	0·059 57	4·771 5	0·209 57
10	4·045 5	0·247 18	20·303	0·049 25	5·018 7	0·199 25
11	4·652 3	0·214 94	24·349	0·041 06	5·233 7	0·191 06
12	5·350 2	0·186 90	29·001	0·034 48	5·420 6	0·184 48
13	6·152 7	0·162 52	34·351	0·029 11	5·583 1	0·179 11
14	7·075 7	0·141 32	40·504	0·024 68	5·724 4	0·174 68
15	8·137 0	0·122 89	47·580	0·021 01	5·847 3	0·171 01
16	9·357 6	0·106 86	55·717	0·017 94	5·954 2	0·167 94
17	10·761 2	0·092 92	65·075	0·015 36	6·047 1	0·165 36
18	12·375 4	0·080 80	75·836	0·013 18	6·127 9	0·163 18
19	14·231 7	0·070 26	88·211	0·011 33	6·198 2	0·161 33
20	16·366 5	0·061 10	102·443	0·009 76	6·259 3	0·159 76
21	18·821 5	0·053 13	118·810	0·008 41	6·312 4	0·158 41
22	21·644 7	0·046 20	137·631	0·007 26	6·358 6	0·157 26
23	24·891 4	0·040 17	159·276	0·006 27	6·398 8	0·156 27
24	28·625 1	0·034 93	184·167	0·005 42	6·433 7	0·155 42
25	32·918 9	0·030 37	212·793	0·004 69	6·464 1	0·154 69
26	37·856 7	0·026 41	245·711	0·004 06	6·490 5	0·154 06
27	43·535 3	0·022 96	283·568	0·003 52	6·513 5	0·153 52
28	50·065 6	0·019 97	327·104	0·003 05	6·533 5	0·153 05
29	57·575 4	0·017 36	377·169	0·002 65	6·550 8	0·152 65
30	66·211 7	0·015 10	434·745	0·002 30	6·565 9	0·152 30
35	133·175 5	0·007 50	881·170	0·001 13	6·616 6	0·151 13
40	267·863 5	0·003 73	1 779·090	0·000 56	6·641 7	0·150 56
45	538·769 2	0·001 85	3 585·128	0·000 27	6·654 2	0·150 27
50	1 083·657 3	0·000 92	7 217·715	0·000 13	6·660 5	0·150 13

Index

Accounting Standards
    Committee 247
accounting year 69
accounts 18, 26, 240-2
acquisition vii, 80-90
  and cash flow 200, 213,
      217-23, 220-2
  financial management 235
  methods ix, 255-6
  policy 4-9
administration 18, 22-4
AETR 181-2
Annual Finance Act 175
asset register 24-6, 46,
      125, 127-8, 197
  and computers 248-9, 252
assets 80, 190
  and cash flow 200, 213,
      216-25
  depreciation 103-5
  financial management
      231-4, 236, 246
  hire companies 21-6
  mortgageable 16

balance sheet 23, 231-3,
      236-7, 241
bankruptcy 107, 199-200
Baxter special price
      indices 111
British Overseas Trade
      Board 261
budget
  administration 22, 190-
      3

capital investment 188
  and cash flow 188-9,
      216-17
  coding system 190-1
  cost 22, 24
  financial 22-3
  forecasts 190-1
  maintenance 124
  master 187-8, 191
  operating 188-90
  purchasing 216-17
  sales 188-90
  transport 188, 190, 192,
      216-17
  workshop 188-90, 192-4,
      216-17
budget control and costing
      x, 187-98, 240
  costing 192-7
  preparation 187-8, 192
  types 188-91

capital 23, 111, 130, 249
  for acquisition 4-8, 10
  borrowed 63-4, 81, 84,
      90
  economic comparisons 46,
      55
  equity 81, 84
  financial management
      231, 233
  fixed return capital
      (FRC) 242
  gearing 242-5
  hiring companies 18-19

investment 188
lock-up 129
shareholders' 84
working 235-40
see also loan capital
capital allowances x, 255,
    231
  and acquisition 80-9
  and cash flow 200, 222-3
  financial management
    228-33, 236-7, 246
  and hire rates 101-2,
    107
  and profitability 69-73
capital cost 31-6, 252
  and acquisition 80, 83
  and hire rates 101, 108,
    110
  and profitability 61-2,
    64, 67, 78
capital investment vi-vii,
    16, 90-1, 265, 270
  and cash flow 221, 225
  and computers 250, 254-5
  financial management
    227, 231, 241-2
  hire rates and 108-9
  in maintenance 122
  in plant ix
  and present worth 30-1,
    33-5
  and profitability 60,
    62, 67, 69-70, 73-5,
    77, 79
capital recovery 31, 34-5,
    58, 66, 108, 112, 270
cash flow x, 81, 107, 199-
    225, 237, 255
  for assets 217-23
  economic comparisons 30-
    3, 35-57
  forecasts 214-19, 222-3,
    237, 240
  from trading operations
    200-18
  interest and time 264,
    269-70
  management 223-5
  and profitability 60-79
cash flow tree 61-2, 68
Companies Act 23, 246-7
computers vii, x, 26, 100,
    264
  basic accounting 250-2

financial appraisals
    254-6
management accounting
    251-4
plant management
    applications 248-56
software packages 100,
    128, 250-2, 255-6
systems 249, 253, 256
Construction Plant Hire
    Association (CPA) 14,
    114, 249
Construction Plant Hire
    Association Conditions
    for Hiring Plant 168,
    170, 172
contribution 201-13, 221
corporate analysis 17-19
corporate trading analysis
    18-19
corporation tax x, 81, 255
  and cash flow 200, 213,
    219-23
  and financial management
    227-30, 232, 236,
    243-5, 247
  and hire rates 101, 107,
    111, 113
  and profitability 69-73
cost 26, 105, 199
  acquisition 5, 84, 86
  capitalised 55-7
  and computers 249, 253
  control 25, 106
  direct, and budgets 189-
    94, 196
  direct, and cash flow
    201-2, 219, 221
  direct, maintenance 124-
    6
  economic comparisons 29,
    49
  financial management
    235, 238
  fixed, and budgets 189,
    192, 197
  fixed, and hire rates
    101, 106, 108-9
  indirect, and budgets
    194, 196
  indirect, and cash flow
    219, 221
  indirect, maintenance
    124, 127

labour 93, 124, 228-9,
    239-40, 251
labour, and budgets 189-
    93, 197
labour, and cash flow
    201-10, 212, 215, 218
operating 20, 30, 44
operating, economic com-
    parisons 57-9
operating, and hire rates
    101, 106-8, 110
operating, and
    profitability 60-2,
    70, 74-5, 77, 79
"packages" 29-30
replacement 252
report 196
running 30-7, 39, 46-7,
    270
running, crane evaluation
    94, 97
running, economic
    comparisons 51, 56,
    58-9
running, and hire rates
    114
and selection 93-4, 97,
    99
variance 192-3
see also capital cost;
    maintenance; owner-
    ship
cost accounting 3, 24
cost centre 192, 195
costing x, 197, 216-17,
    240
credit 199, 223-4
    trade 199-204, 211-17,
    224, 239
credit sale, acquisition by
    81-3
customs 258, 260-1
Customs and Excise 214

de-stocking 211-12, 215
debentures 234, 242-5
depreciation 13, 252
    and budgets 190-4, 196
    and cash flow 219-23
    declining balance 103,
    105, 252
    and financial management
    228-32, 246-7

free 104-5, 107
and hire rates 101-5,
    109, 111-13
overseas 261
sinking fund 103-5
straight line 102, 105,
    108, 252
sum of digits 104-5
DHSS (Department of Health
    and Social Security)
    149
discounted cash flow (DCF)
    yield 60-7, 72-3, 255
    and hire rates 107, 109-
    11, 114
discounting 266, 270
dividend
    and cash flow 200, 213,
    219, 221
    and financial management
    227-8, 230-1, 233,
    243-5
Dixon, J.R. 91-2
DIY small tools vi-7, 12
driver hours 180-2
driver licensing 173, 176-
    8

economic comparisons 29-59
    and inflation 36-46
    present worth 30-6
    replacement age 57-9
    valuation 46-57
EEC directives
    hours law 181-2
    road transport laws 174,
    178
Employers' Liability Act
    (1969) 166
equivalent annual cost
    (EAC) 30, 33-6, 57-9
evaluation
    crane 92-100
    technical 91
Export Group for the Con-
    struction Industries 261

Finance (No.2) Act (1975)
    166, 168
financial accounting
    hire companies 23
    plant division 21-2

financial management x,
    227-47
  accounts 240-2
  balance sheet 227, 231-
    3, 236-7, 241
  capital gearing 242-5
  inflation and 246-7
  plant profitability 245-
    6
  profit and loss account
    227-32
  working capital 235-40

gearing ratio 81, 243
General Agreement on
    Tariffs and Trade
    (GATT) 262

health risks 156-8
health and safety vii, x,
    144-63
  accidents 149-56
  basic provisions 144-8
  check list 151-8
  hire companies 22-3
  legislation 159-63
  policy and organisation
    148-9
Health and Safety at Work
    Act (1974) x, 144-7,
    159, 167, 170
Health and Safety
    Commission 144-5
Health and Safety Executive
    128, 144-5, 147, 149-51
Health and safety
    Inspectorate 145, 147,
    149, 151
heavy goods vehicles 176-
    8, 182
Hire Association of Europe
    (HAE) 14, 114
hire companies ix-x, 12,
    14-20
  and acquisition 5-8, 84-
    5
  and budgets 194, 197
  cash flow 200-1, 214-17,
    219
  and computers 248-50,
    253, 256
  economic comparisons 29-
    30
  financial management
    227-8, 230
  insurance 166, 168-70,
    172
  and maintenance 122,
    127, 130
  organisation 21-6
  overseas operations 261
  and profitability 78-9
  and safety 147
hire purchase 4
  acquisition by 81-3
  and capital investment
    16
  and cash flow 200, 217-
    19, 222-3
  and computers 255
  and financial management
    236-40
  and hire rates 111
  and licensing 179
hire rate 15, 24-6
  and acquisition 5, 7-8,
    13-14, 85
  and budgets 194-5, 197
  calculation 101-15
  and cash flow 216, 221-4
  and computers 249-50,
    252-5
  in economic comparisons
    29, 36
  and profitability 62,
    68-9, 77
hiring vi-vii, 91
  and acquisition 4-10,
    12-26
  acquisition by 80, 84-5
  economic comparisons 36,
    47-8
  and profitability 60-1,
    74, 77

ICE Conditions of Contract
    166-7, 169
import controls 258, 261
inflation x, 193, 225
  and economic comparisons
    30, 36-46
  and financial management
    236, 246-7
  and hiring 13-14, 19,

109, 111-13
and profitability 73-9
inflation index 26
inspection
  hire equipment 24
  and licensing 180-1
  maintenance 120, 122-3,
    127-8, 133
  overseas 258, 261
  pre-shipment 260
inspection x, 26, 148-9,
    153-5, 159-62
insurance vii, x, 22-3,
    26, 83
  benefit 167
  and budgets 192-3
  compulsory 166-7
  construction vehicles
    171
  contingent liability 172
  contractors' all-risk
    policy 168-9, 172-3
  cost 172-3
  employers' liability
    167-8, 172-3
  engineering 167, 169-73,
    225
  and hire rates 101, 105,
    108, 113
  legalities 166-73
  liability 167, 171
  and licensing 174
  and maintenance 127-8,
    133
  material damage 167-71
  motor 167, 170-3
  national 106-7
  overseas 260
  pecuniary 167
  plant and equipment 169-
    70
  public liability 167-8,
    172-3
interest charges 83, 227-
    30, 237, 239, 244
interest and time
    relationships 30,
    264-72

JCT Standard Form of
    Building Contract 166-
    7, 169

Kepner and Tregoe
    (consultants) 91-2

lease
  finance 82-4
  operating 82, 84
lease payments and
    acquisition 83-5, 87,
    90
lease payments and cash
    flow 217, 219-22
leasing 4, 16, 91, 255
  acquisition by 80, 82-90
  and cash flow 200, 217,
    220-1, 223
  financial management
    235-6, 238, 242
licensing vii, x
  driver 173, 176-8
  and hire rates 101, 105,
    108, 113
  import 261
  laws 173-4
  legalities 166, 173-83
  operator 173, 178-82
  public service vehicle
    173
  vehicle excise 173-6
liquidation, voluntary 199
liquidity and cash flow
    199
loan capital 16, 101-2,
    188-9, 199, 230, 247

maintenance vi, x, 3, 26,
    119-43, 148, 254
  and acquisition 5-7, 83-
    4
  and budgets 192-4
  and cash flow 201, 225
  cost 119, 121-5, 130,
    133, 136-42
  financial management
    236, 239
  and hire rates 101-2,
    105-6, 108, 113-14
  and hiring 13-14, 18
  and licensing 179-81
  overseas 258-9
  planned 120-4
  planned corrective 120-2
  planned preventive 120,

122-3
policy 119-23
and profitability 61-2
records 7
and selection 96-7
unplanned 120-1
using computers 248-9
workshop control of 24
maintenance strategy 123-
43
costs 124-7
safety inspections 127-8
stock control 128-43
management
financial 199
hire companies 21-6
manufacturers 111, 200,
215
and maintenance 122-3,
128-9
overseas 107, 258
and safety 146-7
market forecast 14-17
mergers 107

NEDO indices 37
net present value (NPV)
64-5, 67, 73, 88-9

overheads
and acquisition 83, 85
and budgets 192-6
and cash flow 201-2,
213-16, 218-19, 221
and financial management
228-9, 239-40
hire companies 18
and hire rates 101, 106,
108, 113-14
maintenance 124
and profit 25
overseas markets 114, 257-
63
overseas operations vii,
x, 81, 257-63
ownership ix, 3-5, 10,
127, 169
costs 101
costs, and cash flow
201-2, 204, 213, 217,
219, 221-2, 224-5
costs, financial 239-40

economic comparisons 34,
46-9, 53, 57
and hire rates 101, 107-
10, 114
and profitability 60, 77

payback period 67, 72-3,
255
plant
controlled 7-9
division 3, 6, 21, 29-30
holding 6-7
internal subsidiary 8-9
low ownership 8-10
mobile 176
no structure 9
policy 3-10
profitability 245-6
rehire 7-10
selection ix, 5
structures in practice
9-10
present worth 30-6, 109,
255
and acquisition 81, 85-9
economic comparisons 37,
39-44, 46-9, 51-9
interest and time 265-6,
268-70
and profitability 63-7,
74-5, 77
present worth factors 86-
9, 109
economic comparisons 31-
2, 39-44, 47-8, 51-5,
58
interest and time 265-6,
268-9
and profitability 65-6,
73-5, 78
profit 16, 91
and acquisition 7-8, 83
assessable 228-9
budgeted 193-7
and cash flow 199, 201-
2, 213, 219, 221-3
and computers 253, 255-6
financial management and
227, 230, 238, 242-7
hire companies 14, 19,
25
and hire rates 102, 107-

9, 111, 113
pre-tax   230
profit centre   x, 5-6, 9-10
profit flow   80-1, 83-4,
    87-90
profit and loss account
    23, 107, 187-8, 227-32
profitability   13, 199
  and acquisition   6, 8-10,
      80
  average annual rate of
      return   67, 255
  and budgets   188, 196-7
  and computers   248, 255
  and financial management
      229, 240-1
  hire companies   15-16,
      19, 25-6
  and hire rates   107
  of maintenance   121
  measuring   60-7
  plant   245-6
  of plant division   3
purchase
  and acquisition   7-9
  acquisition by   80-90
  and cash flow   200, 215,
      217-23
  and computers   249, 252,
      254
  credit   217
  and depreciation   13
  economic comparisons   34,
      36-7, 39, 44, 46, 48,
      51-2, 54-5, 57-9
  financial management
      228-9, 232, 234, 246
  and hire rates   101, 103,
      106-8, 111-12, 114
  hire and rental   12
  interest and time   270
  and leasing   83
  and maintenance   123,
      125, 138
  outright   81-2, 217, 219,
      222-3, 238, 255
  and profitability   60-2,
      70-1, 74
  and selection   91-2, 94,
      97
  straight   4
purchasing
  hire companies   23-4
  plant division   21-2

rate of return
  apparent   76-8, 112
  real   76-8, 112
recession   4, 13-14, 77,
    114-15
Registrar of Companies   227
rehiring   7-9
reinvestment   227-8
rental   vi, x, 7, 12-26,
    83, 235
rental companies   14-16,
    21-6, 25
rental rates   19
replacement   252
  age   57, 78, 267
  economic comparisons   31,
      36, 49-59
  and hire rates   111-12
  and maintenance   119-21,
      123
resale   9, 36-7, 44
  economic comparisons   48,
      51, 53-6
  and hire rates   102
  interest and time   264
  and profitability   60-2
resale value   32-3, 35
  and acquisition   81, 83
  economic comparisons   46-
      7, 56-9
  and hire rates   108, 110-
      12, 114
  and profitability   60-2,
      70-2
retail price index   37
Road Traffic Act (1972)
    166, 171-2, 176
Road Traffic Act (1974)
    178
road transport legislation
    173-4

safety   94, 97, 121
  see also health and
      safety
safety check list   151-63
safety regulations   22-3,
    119, 144-63
safety representatives
    149, 151
safety standards   120-1
safety statistics   147-8
scrap   152

scrap value   102-4
selection   37, 91-100
sensitivity analysis   68-9
service centre   x, 6, 9-10,
    14
spreadsheet   100, 250
stock   19
    financial management
        228-9, 231, 233-4,
        238-9, 241, 246-7
    overseas   258
    safety level   131-3
stock control   24, 128-43
    ABC method   129-30, 143
    and computers   248-51,
        254
    and continuous usage
        140-3
    discounts   138-40
    economic order quantity
        133-43
    inventory control   130-43
    shortages   135-8
stocks   201, 206-13, 215-
    17, 222-4

Trade and Industry,
    Department of   227
trade union   149
trading analysis   19
training   22-3, 123
transport   25, 218
    financial management   239
    and hire rates   101, 108-
        9
    and maintenance   128-9
    overseas   257-61
    and safety   154

Transport Acts   178, 181
turnover   200, 228, 230
    and financial management
        235-6, 242, 245-6
    hire company   12, 15, 18-
        19
    of plant structures   9-10

utilisation   ix, 109, 127
    and acquisition   6-7, 85
    and computers   249, 253,
        255
    and profitability   61-2,
        68
    reports   194-5, 197, 253
    variance   195-6
utilisation levels   4, 8-
        10, 25-6, 119, 121
    and hiring   13, 19, 102

valuation, equipment   46-57
variance   192-3, 253
    analysis   194-6, 248
    price and sales   194-6
    utilisation   195-6
VAT (value added tax)   200-
    1, 213-14, 219, 221
Vehicles (Excise) Act
    (1971)   174-7

wages   22-5, 101, 106-7,
    111
Weights and Measures Act
    (1985)   174
wordprocessing   250
workshop control   24-5